Pharmaceutical Discovery, Development and Manufacturing Forum 2022

Held at the 2022 AIChE Annual Meeting

Phoenix, Arizona, USA
13-18 November 2022

ISBN: 978-1-7138-7897-1

Printed from e-media with permission by:

Curran Associates, Inc.
57 Morehouse Lane
Red Hook, NY 12571

Some format issues inherent in the e-media version may also appear in this print version.

Printed with permission by Curran Associates, Inc. (2023)

AIChE
120 Wall Street, FL 23
New York, NY 10005-4020

Phone: (800) 242-4363
Fax: (203) 775-5177

www.aiche.org

Additional copies of this publication are available from:

Curran Associates, Inc.
57 Morehouse Lane
Red Hook, NY 12571 USA
Phone: 845-758-0400
Fax: 845-758-2633
Email: curran@proceedings.com
Web: www.proceedings.com

TABLE OF CONTENTS

ADVANCES IN NEW MODALITIES: PREDICTIVE MODELING TECHNOLOGIES

APPLIED FORMULATION DESIGN IN DRUG PRODUCT

COMPUTATIONAL SOLID STATE PHARMACEUTICS I

CONTINUOUS PROCESSING IN DRUG SUBSTANCE: MODELLING AND SIMULATION

CONTROL STRATEGIES IN PHARMACEUTICAL DEVELOPMENT AND MANUFACTURING I

ENABLING TECHNOLOGIES: DRUG SUBSTANCE AND DRUG PRODUCT MANUFACTURING

INTEGRATED PRODUCT AND PROCESS DESIGN WITH PHARMACEUTICAL APPLICATIONS I

PHARMA 4.0 (ADVANCED CONTROLS, PROCESS AUTOMATION, DATA ANALYTICS, ETC.) IN DRUG SUBSTANCE AND DRUG PRODUCT I

PREDICTIVE SCALE-UP/SCALE-DOWN FOR PRODUCTION OF PHARMACEUTICALS AND BIOPHARMACEUTICALS I

PANEL SESSION: PRE-COMPETITIVE COLLABORATIONS THROUGH THE ENABLING TECHNOLOGIES CONSORTIUM (ETC)

FACILITIES OF THE FUTURE (INVITED TALKS)

ADVANCEMENTS IN PARTICLE ENGINEERING AND MATERIAL SCIENCES IN PHARMACEUTICAL PROCESS DEVELOPMENT II

ADVANCES IN DRUG DISCOVERY PROCESSES (INCLUDING HTE): PROTEIN ENGINEERING APPROACHES WITH TARGET THERAPEUTIC APPLICATIONS

CONTROL STRATEGIES IN PHARMACEUTICAL DEVELOPMENT AND MANUFACTURING II

INTEGRATED PRODUCT AND PROCESS DESIGN WITH PHARMACEUTICAL APPLICATIONS II

PHARMA 4.0 (ADVANCED CONTROLS, PROCESS AUTOMATION, DATA ANALYTICS, ETC.) IN DRUG SUBSTANCE AND DRUG PRODUCT II

PREDICTIVE SCALE-UP/SCALE-DOWN FOR PRODUCTION OF PHARMACEUTICALS AND BIOPHARMACEUTICALS II

CONTINUOUS PROCESSING IN DRUG SUBSTANCE: ADVANCEMENTS IN INDUSTRY

CONTINUOUS PROCESSING IN DRUG SUBSTANCE AND DRUG PRODUCT: INTEGRATED PROCESSES

ENABLING TECHNOLOGIES: PROGRESS IN TOOLS AND TECHNOLOGIES

ENABLING TECHNOLOGIES: MECHANISTIC AND STATISTICAL MODELING

ADVANCES IN NEW MODALITIES: BIOLOGICS AND LARGE MOLECULES

ADVANCES IN NEW MODALITIES: PEPTIDES, NUCLEIC ACIDS

COMPUTATIONAL SOLID STATE PHARMACEUTICS II

MODELING SOLUBILITY, DISSOLUTION, PERMEABILITY AND DRUG DELIVERY

COMPUTATIONAL APPROACHES TO DOE AND BETTER PROCESS UNDERSTANDING

672a Control of Crystal Morphology through Process Parameter Optimization Guided By 2d PBM

Brian Seifried, Tamar Rosenbaum, John Gamble, Joshua Engstrom, Ulrich Schacht, Yash Melkeri, Niall Mitchell

The primary method of controlling API powder properties is through the design of a robust API crystallization process. The crystallization process has historically been developed through a combination of screening work and design of experiments (DoE), which can be both materials and labor intensive. To rapidly address evolving API powder properties control needs, we have developed a hybrid modeling and experimental workflow for crystallization process development. The workflow involves creation of a population balance model (PBM) via gPROMS FormulatedProducts (gFP, PSE Siemens), based on desupersaturation and PSD data gathered from initial crystallization experiments utilizing EasySampler equipped with a novel filter frit (Mettler Toledo), and Morphologi G4 (Malvern Panalytical Ltd). We demonstrate the effectiveness of the workflow by applying it to a needle-shaped API (Figure 1,b) with a cooling crystallization process. Separate growth kinetics were used to describe the major and minor axis growth, and it was found necessary to utilize a 2D seed model to accurately describe the seed PSD. The model provided a good fit to the desupersaturation data and particle aspect ratio (AR), accurately capturing that the relative supersaturation (RSS) at seed point was a main factor in influencing product AR (Figure 1, a). The model was then used to probe product AR when the seed type (PSD and AR) was varied. The model predicted that use of lower AR seeds would yield lower product AR that in turn led to efforts to generate more equant seeds both via micronization (jet milling) and temperature cycling. Experiments with the different seed types validated the model predictions and yielded lower AR products (Figure 1, c).

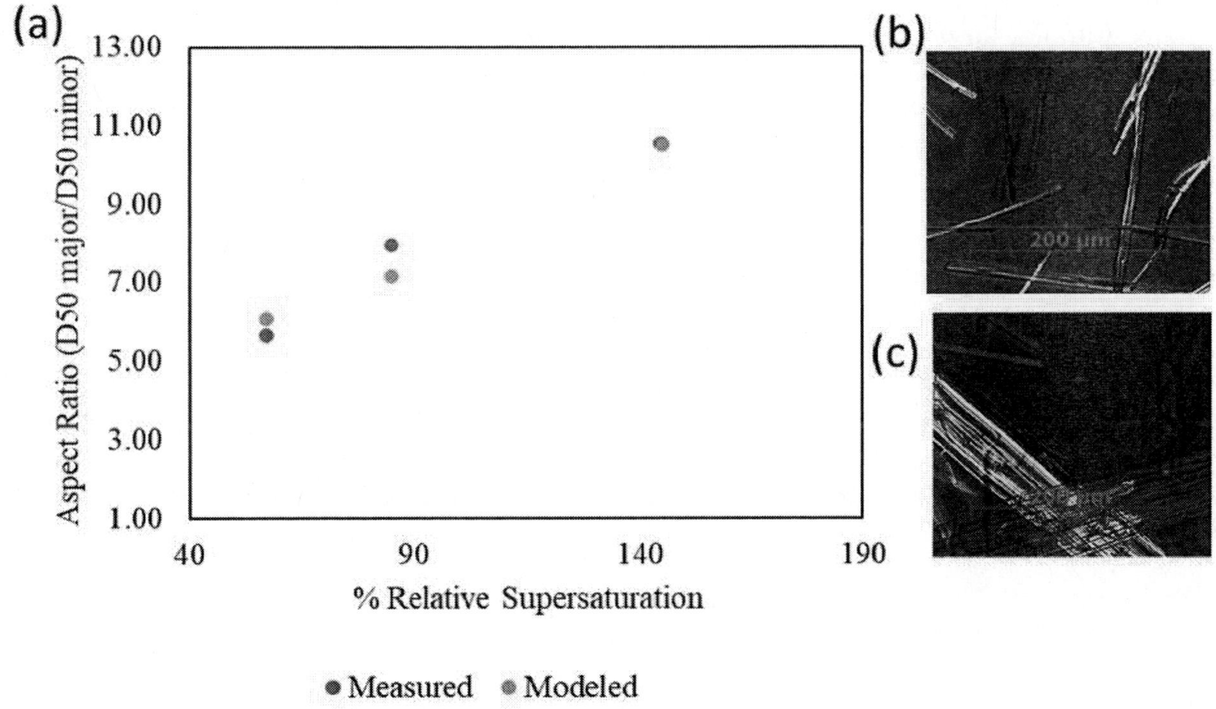

Figure 1. (a) Comparison of measured and modeled aspect ratio vs % RSS at seed point, (b) optical microscope image of slurry showing high-aspect ratio needles grown in high RSS (c) optical microscope image of slurry showing thicker rods when grown in low RSS

672b Introducing Contcarsim, a Benchmark Simulator for Design and Control of Continuous Integrated Filtration-Drying of Crystallization Slurries

Francesco Destro, Massimiliano Barolo, Zoltan Nagy

Within the recent modernization momentum in the pharmaceutical industry, the transition to continuous manufacturing and the introduction of closed-loop control on quality have been identified as key enablers for delivering the next generation of pharmaceutical quality (Fisher et al., 2016; Destro et al., 2022). However, their practical adoption in industrial plants is still lagging behind, either because of the economic cost for their implementation, of the lack of personnel trained for this purpose, or even because of concerns about the regulatory approval of novel technology (Grangeia et al., 2020; Reinhardt et al., 2020).

In this presentation, we demonstrate advanced operation design, monitoring and control applications in pharmaceutical development and manufacturing through ContCarsim, a novel simulator for continuous integrated filtration-drying of crystallization slurries. ContCarSim, publicly released in the MATLAB environment, aims at promoting the adoption of advanced operation design, process monitoring and control by the pharmaceutical community within the quality-by-design and quality-by-control initiatives (ICH, 2009; Su et al., 2017), by making available a digital environment where the user can easily test and benchmark control strategies, or generate data for process analytics studies.

ContCarSim allows the user to simulate the operation of a novel carousel technology for continuous integrated filtration-drying, in normal operating conditions or under a set of different disturbance scenarios. Furthermore, the user-friendly structure of ContCarSim allows the implementation on the simulator of user-conceived control strategies, which may include advanced elements such as, for instance, model predictive control and state estimation. We demonstrate the use of ContCarSim through the implementation of a three-layer control system, featuring an end-point controller based on state estimation, which terminates filtration and drying cycles when the target product quality has been achieved. The proposed control system also includes a real time optimization routine, for maximizing the throughput of dry crystals obtained from the unit operation.

References

Destro, F., Hur, I., Wang, V., Abdi, M., Feng, X., Wood, E., Coleman, S., Firth, P., Barton, A., Barolo, M. and Nagy, Z.K., 2021. Mathematical modeling and digital design of an intensified filtration-washing-drying unit for pharmaceutical continuous manufacturing. *Chemical Engineering Science*, *244*, p.116803.

Destro, F., Barolo, M., 2022. A review on the modernization of pharmaceutical development and manufacturing-Trends, perspectives, and the role of mathematical modeling. *Int. J. Pharm.*, p.121715.

Fisher, A.C., Lee, S.L., Harris, D.P., Buhse, L., Kozlowski, S., Yu, L., Kopcha, M., Woodcock, J. (2016). Advancing pharmaceutical quality: An overview of science and research in the U.S. FDA's Office of Pharmaceutical Quality. *Int. J. Pharm.* **515**, 390–402.

Grangeia, H.B., Silva, C., Simões, S.P., Reis, M.S. (2020). Quality by design in pharmaceutical manufacturing: A systematic review of current status, challenges and future perspectives. *Eur. J. Pharm. Biopharm.* **147**, 19–37.

ICH (2009). Q8(R2) - Pharmaceutical development. *Q8(R2)*. Available at: https://www.fda.gov/media/71535/download

Reinhardt, I.C., Oliveira, J.C., Ring, D.T. (2020). Current perspectives on the development of Industry 4.0 in the pharmaceutical sector. *J. Ind. Inf. Integr.* **18**, 100131.

Su, Q., Moreno, M., Giridhar, A., Reklaitis, G. V., Nagy, Z.K. (2017). A Systematic Framework for Process Control Design and Risk Analysis in Continuous Pharmaceutical Solid-Dosage Manufacturing. *J. Pharm. Innov.* **12**, 327–346.

672c Equipment Optimization for Cryogenic Particle Formation Correlating Experimental and Model Results

Howard Stamato, Nima Yazdanpanah, Jorge Lopez, Russel Carson, Edwin Amoro, Elliott Bay, Jason Antunovich

Cryogenic particle formation is applicable across industries and is often a first step to further drying at low pressure and temperature for a stable solid form of otherwise vulnerable materials (1,2,3,4). This essential process step is governed by complex phenomena where, in some cases, the actual process performance is still influenced by the machinist's skill. In this abstract the effective use of an in-silico model delivered in the short term to inform and expedite physical models and experimentation is described.

Rapid particle formation at cryogenic conditions includes complex phenomena. The complexity is much more severe when the consistent particle properties (such as size, homogeneity, microstructure) are required for drug delivery and medical applications, and in addition, when the solution has nonnewtonian behavior, droplet generation and particle formation is fast, thus multiscale, multiphysics models are involved. The equipment, process, and product design for these cases can be extremely costly, time consuming, and inefficient, due to the complexities, fast transport phenomena, and nonlinearities.

The trial and error experimental efforts have been carried out conventionally for many cases and products. However, this approach is not comprehensive and less affordable in the long term. The value of in-silico modeling and first-principle analysis is well established and grows with every advance in software and computing platforms (5,6).

In this work the effective use of a high-fidelity simulation delivered in a short time to inform physical modifications and testing is described. The nozzles of a cryogenic particle formation machine were modified both in-silico and validated experimentally to understand and improve droplet formation prior to freezing. The process is used in production of powders for inhalation (7).

In this presentation, the simulations, underlying physics, experimental validation, model implementation, and industrial scale application will be discussed. The parametric study for different variables, sensitivity analysis, computation fluid dynamic (CFD) for Plateau–Rayleigh instability, and pressure wave resonance in droplet generation will be discussed in detail along with lab measurement and experimental data validation.

References:

1. D Fissore Freeze-drying in the coffee industry, 2015 https://www.newfoodmagazine.com/article/16968/freeze-drying-in-the-coffee-industry/ Accessed March 19, 2022
2. TL Rogers, KP Johnston, RO Williams III, Solution-Based Particle Formation of Pharmaceutical Powders by Supercritical or Compressed Fluid CO2 and Cryogenic Spray Freezing Technologies, Drug Development and Industrial Pharmacy 27(10) 2001, pp 1003-1015.
3. H Shiffter, J Condliffe, S Conhoff, Spray-freeze-drying of nanosuspensions: the manufacture of insulin particles for needle-free ballistic powder delivery, Journal of the Royal Society Interface 2010 Aug 6; 7 Suppl 4 S483-500

4. Droplet Stream Freeze-Drying for the Production of Protein Formulations: From Simulation to Production S Wanning, M Jager, SN Eggerstedt, R Suverkrup, A Lamprecht, M Dietzel, M Kuschel, A Darvan, M Sommerfeld, in Process Spray U Fritsching, (ed) Springer 2 August 2016, pp.341-381
5. E C Joseph, S Conway, C Dekate, Economic Model For a Return on Investment Analysis of United States Government High Performance Computing (HPC) Research and Development (R andamp; D) Investment, Technical Report for U. S. Department of Energy Office of Science and Technical Information September 2013 https://www.osti.gov/biblio/1156830 accessed March 19, 2022
6. J Kostal A Voutchkova-Kostal Going All In: A strategic Investment in In Silico Toxicology, Chem Res. Toxicol. 2020, 33(4), 880-888
7. Amoro, K Vanackere, MA White Apparatus and method for cryogranulating a pharmaceutical composition U.S. Patent 8950320B2

672e Development of a Crystallization Process to Prevent Oiling and Optimize Physical Properties of a Drug Substance

Peter Galebach, Zachary Garlets, Andrew Werneth, Melda Sezen-Edmonds

The need for new medications has necessitated increasing complexity in the molecular structure of modern pharmaceutical drug substances. As these molecules become more complex, isolating them in crystalline form becomes increasingly difficult due to challenges such as liquid-liquid phase separation (LLPS), poor desaturation kinetics, and poor particle morphology. These challenges can have negative consequences for manufacturing processes at scale resulting in inconsistent operational and quality attributes including: impurity rejection, powder properties, and crystal form. In this work we discuss an industrial case study for a challenging active pharmaceutical ingredient (API) crystallization which exhibits LLPS, slow desaturation kinetics, poor powder properties, and inconsistent form.

We describe development of the crystallization process using a comprehensive experimental and modeling approach which includes: solubility and salt screening, upstream process modification, and crystallization parameter optimization (addition rates, seed loading, temperature cycling, etc.). LLPS of the API could be avoided through proper pH control before crystallization which was verified by modeling of the pKa of various basic sites on the compound and corresponding modification of upstream processing. Desaturation kinetics and morphology of the compound was tracked with PAT (FBRM, IR) which was utilized to model the kinetics of the crystallization. Modeling showed slow growth and that nucleation accounted for most desaturation. We optimized an unseeded crystallization by varying anti-solvent addition rates and temperature cycling to tune the physical properties of the drug substance while maintaining high yield. Finally, we utilized a recrystallization to achieve consistent crystal form.

672f Investigating the Effect of Droplet Confinement and Porous Confinement on the Formation of Carbamazepine Polymorphs

Alice Parkes, Ahmad Ziaee, Gavin Walker, Emmet O'Reilly

Controlling crystal polymorphism is one of the greatest challenges in the development of active pharmaceutical ingredients (APIs). Polymorphs of an API are different structural arrangements of the same molecule in a crystal lattice [1]. Each polymorph has different physicochemical properties such as: stability, solubility, and bioavailability [2]. Moreover, crystal polymorphism has historically brought about a lot of issues in drug development and patient safety; well-known cases include HIV-1 drug, Ritonavir, and antacid drug, Zantac [3-5]. As a result, control over polymorphism in the pharmaceutical industry has been highly sought after.

This study outlines a systematic one-step approach to controlling the polymorphism of carbamazepine (CBZ) via droplet confinement using spray drying as a continuous method of manufacture. In this process, CBZ molecules are confined within spray dried droplets of varying sizes during the nucleation process to investigate whether different levels of confinement result in obtaining different polymorphic forms. Spray drying is a continuous method for converting a liquid feed solution or suspension to powder in a single step [6]. Different atomising gas flow rates were used to vary the size of the droplet produced in the spray dryer. As a result, a trend emerged in the polymorphic form of the spray dried powders with respect to the sprayed droplet sizes. PXRD confirmed the largest droplet size (~38.55 µm) produced both the stable polymorph, form III, and metastable form IV, whereas only the metastable polymorph, pure form IV, was isolated from the smallest droplet size (~5.39 µm). Accelerated stability tests revealed that smaller level of confinement yields particles with higher stability. The amount of the most stable polymorph of CBZ, form III, present in the particles produced by the larger droplets increased during the stability test. However, the particles produced by the smaller droplets, remained as pure metastable CBZ form IV throughout the stability study. To the best of our knowledge, this is the first study to successfully obtain pure CBZ metastable form IV in bulk using a scalable and controlled method of continuous manufacture.

Analytical techniques used in this study include particle size analysis (PSA) (fig.1A), Powder x-ray Diffraction (PXRD) (fig.1B), Scanning Electron Microscopy (SEM), Differential Scanning Calorimetry (DSC) and Fourier-Transform Infrared (FTIR) Spectroscopy.

As a follow up investigation, we conducted a systematic approach to controlling the polymorphism of CBZ via solid confinement within porous silica using spray drying and rotary evaporation. Silica is a biocompatible material which has the potential to stabilise metastable polymorphic forms of APIs with enhanced dissolution properties [7]. This hypothesis-driven investigation was carried out to explore the transferability of the concept that the size of a pore must be 20 times the molecular radius of a compound for controlled crystallisation to occur [8, 9]. Carbamazepine molecules were confined within silica pores of varying size during the nucleation process to investigate the effects of porous confinement size on its solid-state and the final polymorphic form. The critical size for crystallisation of CBZ in confinement is 20 times its molecular radius, 9.8 Å [10]. In order to clearly understand the effect of confinement, three silica samples were used in this study;-one with a pore size smaller than the critical size for crystallisation to occur, a second with a pore size close to the critical size, and a third with a pore size larger than the critical size. Molecular dynamic simulations were employed as part of this investigation to predict the behaviour of different numbers of molecules confined within different pore sizes.

These methods of controlling the polymorphism of CBZ by different methods of confinement and stabilising metastable polymorphs through continuous methods of manufacture can have significant implications for controlling the polymorphism of other polymorphic APIs, cocrystals and multicomponent pharmaceuticals.

References:

1. Cruz-Cabeza, A.J., N. Feeder, and R.J. Davey, *Open questions in organic crystal polymorphism.* Communications Chemistry, 2020. **3**(1): p. 142.

2. Vippagunta, S.R., H.G. Brittain, and D.J.W. Grant, *Crystalline solids.* Advanced Drug Delivery Reviews, 2001. **48**(1): p. 3-26.

3. Bauer, J., et al., *Ritonavir: An Extraordinary Example of Conformational Polymorphism.* Pharmaceutical Research, 2001. **18**(6): p. 859-866.

4. Bernstein, J., *Polymorphism in Molecular Crystals 2e.* Vol. 30. 2020: International Union of Crystal.

5. Desiraju, G.R., *Crystal gazing: structure prediction and polymorphism.* Science, 1997. **278**(5337): p. 404-405.

6. Ziaee, A., et al., *Spray drying of pharmaceuticals and biopharmaceuticals: Critical parameters and experimental process optimization approaches.* European Journal of Pharmaceutical Sciences, 2019. **127**: p. 300-318.

7. Gao, Y., et al., *Multifunctional Role of Silica in Pharmaceutical Formulations.* AAPS PharmSciTech, 2022. **23**(4): p. 90.

Jones, E.C.L. and L.M. Bimbo, *Crystallisation Behaviour of Pharmaceutical Compounds Confined within Mesoporous Silicon.* Pharmaceutics, 2020. **12**(3).

9. Sliwinska-Bartkowiak, M., et al., *Freezing behavior in porous glasses and MCM-41.* Colloids and Surfaces A: Physicochemical and Engineering Aspects, 2001. **187-188**: p. 523-529.

10. Suriyanon, N., P. Punyapalakul, and C. Ngamcharussrivichai, *Mechanistic study of diclofenac and carbamazepine adsorption on functionalized silica-based porous materials.* Chemical Engineering Journal, 2013. **214**: p. 208-218.

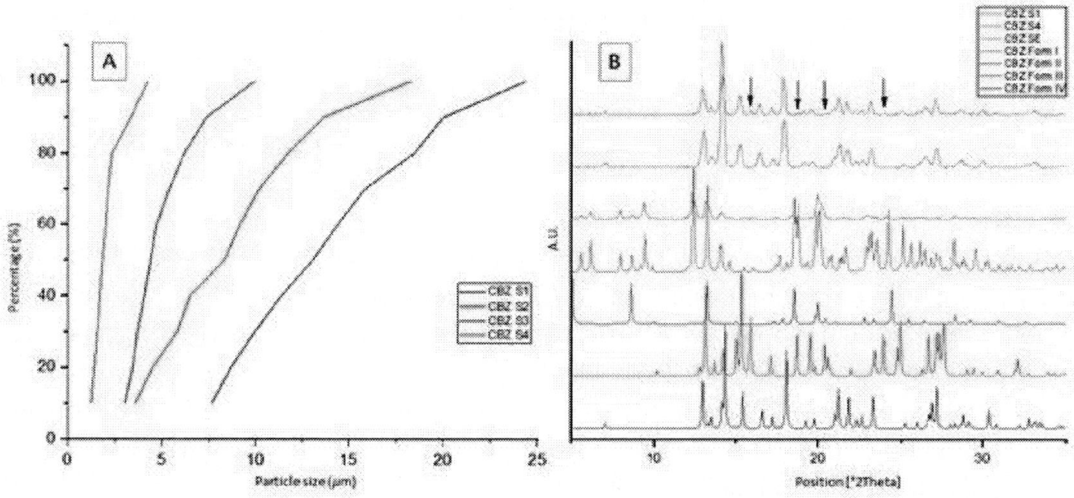

Figure 1 (A) Particle size variation for 100 particles produced by each droplet size (B) PXRD graph of CBZ S1, CBZ S4, CBZ slow evaporation (SE) sample and CBZ polymorphic forms I, II, III, and IV from literature which shows the effect of particle size on the polymorphic form of CBZ.

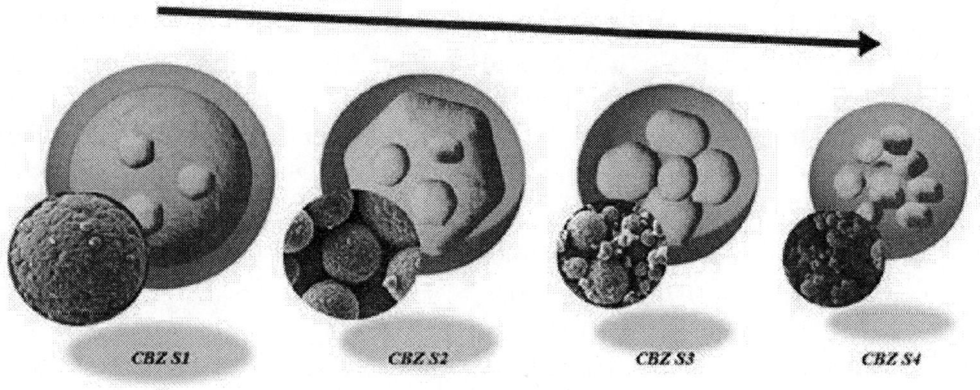

672g Engineering Theranostic Superparamagnetic Nanoparticles for Hyperthermia and Magnetic Particle Imaging Using a Quality-By-Design Approach

Shaquib Rahman Ansari, Ioannis Katsaros, Michelle Åhlén, Christel A.S. Bergström, Peter Svedlindh, Carlos Rinaldi-Ramos, Alexandra Teleki

Image guided therapy integrating magnetic particle imaging (MPI) and magnetic hyperthermia (MH) is of great potential interest as it enables personalized and precise treatment.[1] The properties of superparamagnetic iron oxide nanoparticles (SPIONs) to release heat in an alternating magnetic field (AMF) and to act as contrast agents make them an excellent theranostic modality. The development of tailored SPIONs is paramount to achieve high sensitivity and good resolution for imaging, and efficient and controlled hyperthermia.[2] Therefore, to support the clinical translation of SPIONs, here we apply a quality-by-design (QbD) approach to systematically investigate the effect of SPION particle properties on their MPI and MH performance.

SPIONs were produced by scalable flame spray pyrolysis (FSP)[3] technique. The QbD approach, implemented risk analysis and design of experiments (DoE) to link the SPION properties with MPI and MH performance. First, the effects of FSP process parameters were correlated with the SPION crystal size using a factorial design. This DoE was modelled using multiple linear regression (R^2 0.98, Q^2 0.95). The nanoparticle crystal size was strongly affected by the precursor concentration and flow rate, and the dispersion gas flow rate. Based on this DoE model, the nanoparticle size and surface area can be fine-tuned during their manufacturing. Then, a subsequent DoE linked the crystal size and composition of SPIONs with their saturation magnetization (M_s), intrinsic loss power (ILP) and MPI performance. The crystal size was varied between 6 to 30 nm (to yield superparamagnetic particles) and the SPION composition ($M_xFe_{3-x}O_4$) was modified using four dopants (M = Zn, Co, Mn, Mg) at three doping concentrations (x = 0.25, 0.5, 0.75).

Modelling of the DoE showed that the choice of dopant has a significant effect on the M_s. Doping with Co, Mn and Zn increased the M_s, whereas Mg doping significantly decreased it. Furthermore, larger crystal size and lower doping content was strongly linked to an increase in M_s. Preliminary studies of the hyperthermia performance, as assessed by ILP, also indicated a strong correlation with the dopant. Manganese showed the highest ILP amongst all the dopants. Moreover, the ILP was found to be higher in SPIONs > 13 nm compared to smaller ones.

Overall, MPI and MH performance of SPIONs show strong relationship with their size and composition. This study demonstrates a systematic QbD approach to develop theranostic SPION-based drug delivery system and supports their early development towards clinical applications.

Acknowledgement: This project has received funding from the European Research Council (ERC) under the European Union's Horizon 2020 research and innovation programme (grant agreement No. 101002582). The authors also acknowledge financial support from the Science for Life Laboratory.

References:

(1) Dadfar, S. M.; Roemhild, K.; Drude, N. I.; von Stillfried, S.; Knüchel, R.; Kiessling, F.; Lammers, T. Iron Oxide Nanoparticles: Diagnostic, Therapeutic and Theranostic Applications. *Adv. Drug Deliv.*

Rev. **2019**, *138*, 302–325.

(2) Du, Y.; Liu, X.; Liang, Q.; Liang, X.-J.; Tian, J. Optimization and Design of Magnetic Ferrite Nanoparticles with Uniform Tumor Distribution for Highly Sensitive MRI/MPI Performance and Improved Magnetic Hyperthermia Therapy. *Nano Lett.* **2019**, *19* (6), 3618–3626.

(3) Mueller, R.; Mädler, L.; Pratsinis, S. E. Nanoparticle Synthesis at High Production Rates by Flame Spray Pyrolysis. *Chem. Eng. Sci.* **2003**, *58* (10), 1969–1976.

565a CDER Efforts on Pharmaceutical Manufacturing Modernization

Stelios C. Tsinontides

The presentation will start with a brief introduction of the concept of Pharmaceutical Quality, and then will describe CDER's Quality journey from the early 2000s to the present to promote use of Advanced Technologies in bio-pharmaceutical manufacturing to achieve a higher level of process capability and thus higher product quality assurance. The presentation will discuss most recent efforts at the Agency, internal and external, to develop streamlined processes to facilitate approval of applications implementing new technologies, and to engage with industry and other external stakeholders to solicit their input on how to promote quality manufacturing. These efforts include, but not limited to, implementation of IQA teams and KASA in the assessment process, the Emerging Technologies Program (ETP) and the Framework for Regulatory Advanced Manufacturing Evaluation (FRAME) to facilitate a close collaboration between Industry and Regulators to bring such advanced innovative technologies to bear to enhance product quality assurance. Establishing an appropriate regulatory framework is essential to keep pace with innovation to enable approval of critical and high-quality medicines in a timely and cost-effective fashion by increasing the agility, efficiency and flexibility of pharmaceutical manufacturing and supply chains.

565b Molecular Discovery with Limited Human Input Using a Machine Learning Guided Automated Platform

Matthew McDonald, Brent Koscher, Richard Canty, Seung Kyun Ha, Camille Bilodeau, Klavs Jensen

Discovery of new small molecules often involves synthesizing and testing a rationally designed family of molecules based on a known biologically active motif. Recently, machine learning methods have led to drug discoveries by repurposing known/old compounds [1] and generating new molecules with targeted protein interactions [2]. Candidate small molecules should be moderately lipophilic and relatively non-toxic, as well as biologically potent. We have developed a well-plate-based platform for the automated optimization of molecular scaffolds across multiple properties. Machine learning property models are coupled with a predictive retrosynthetic model (ASKCOS) and robotic modules to suggest, synthesize, and test candidate molecules. The discovery process is iterative, with initial synthesis campaigns being used to retrain and improve property model predictions for a high-performer iteration. The properties used to demonstrate the platform are water/octanol partitioning [3], cytotoxicity based on HEK293T cells, and optical spectral properties [4], as a stand-in for plate-reader-based biological activity assays.

This talk will focus on efforts to improve automation to enable the entire discovery cycle—initial explorative target selection, synthesis, characterization, model refining/retraining, and best-performer synthesis—to be executed with limited human interaction. Liquid handling, reaction optimization, air-free synthesis, high-temperature synthesis, workup, and isolation, all in a well plate format, have been automated and interfaced to the ML models. This combined approach to small molecule exploration, where several properties are optimized automatically and simultaneously, could allow medicinal chemists to focus on scaffold/family discovery rather than more tedious scoping experiments. If evaluation of whole families of compounds were entirely automated, more effort could be applied to identifying targets for small molecule drugs.

[1] J. M. Stokes, et al., *Cell*, 2020, **180**, 688-702.e613.

[2] Y. Bian and X.-Q. Xie, *Journal of Molecular Modeling*, 2021, **27**, 71.

[3] F. H. Vermeire and W. H. Green, *Chemical Engineering Journal*, 2021, **418**, 129307.

[4] K. P. Greenman, W. H. Green and R. Gomez-Bombarelli, *Chemical Science*, 2022.

565c Computational Discovery of Peptide Inhibitors That Neutralize *C. Diff.* Toxin a in Jejunum and Colon Epithelial Cells

Sudeep Sarma, Xingqing Xiao, Stefano Menegatti, Nathan Crook, Scott Magness, Carol Hall

Clostridium difficile infection is the leading cause of diarrhea and colitis (inflammation of the colon) in North America and Europe. The fraction of the population infected by the disease is increasing as new strains associated with significant morbidity and mortality have appeared. There is a growing concern about the failure of the first line of treatment for *C. diff.* infection (metronidazole and vancomycin) and thus, current attention is focused on the need for alternative treatment options like biologic drugs which are safer and more efficacious. A viable and cost-effective strategy is the development of targeted peptide-based therapeutics that prevent and treat *C. diff.* infections by deactivating the pathogen while leaving the resident gut microbiota unharmed.

During infection, the *C. diff.* pathogen produces two large virulent toxins (toxins A and B) that share 71% sequence homology. The glucosyltransferase domain (GTD) of these toxins acts by binding Uridine diphosphate (UDP)-glucose, hydrolyzing it into glucose and UDP, and attaching the glucose monomer to the human Rho family of GTPases. Glycosylation of the GTPases by toxin A and toxin B GTDs leads to disruption of the cytoskeleton, apoptosis, and death of the colon epithelial mammalian cells.

The objective of this project is to computationally design peptide inhibitors that bind with high affinity and specificity to *C. diff.* toxin A GTD to inhibit its activity. We use an automated ***Pep****tide* ***B****inding* ***D****esign* (PepBD) algorithm developed in our lab to design peptide sequences that bind to *C. diff.* Toxin A. Our peptide search algorithm employs Monte Carlo methods, self-consistent mean-field theory, and the concerted rotation (CONROT) technique to search for peptides in sequence and conformation space that bind to a target protein. The starting input structure for the algorithm requires a reference ligand, which is usually an experimentally determined peptide sequence that has a good binding affinity to the target protein.

We initially discovered a 10-mer peptide inhibitor, NPA, using EGWHAHTGGG (K_d = 100 nM) as the reference ligand, a peptide identified by the Feig lab using phage display. The peptide showed good neutralization against toxin A when tested *in-vitro* in jejunum (intestinal) cells but showed no effect in colon cells. Based on feedback from experiments and further computational analysis we designed shorter peptide inhibitors. These peptides neutralize toxin A in both jejunum and colon cells. Current work is focused on furthering our understanding of the interaction of these peptides with the toxins, and discovering peptides that can bind to Toxin B.

565d Identification of Resistance Determinants in Laboratory Evolved *Mycobacterium Smegmatis* evolved Under Temporally Variable Drug Profiles

Akanksha Kashyap, Sarika Mehra

Antibiotic resistance continues to be a global health concern with emergence of resistance in bacteria out-competing the antibiotic discovery rate. Adaptive Laboratory Evolution (ALE) is a useful approach to investigate factors that govern the evolution of resistance in bacteria in order to identify strategies to combat them.

In our work, we have studied the impact of temporally varying environments on resistance evolution in *Mycobacterium smegmatis* subjected to a range of constant and increasing antibiotic concentrations. Parallel adaptive evolution was carried out for a range of constant and step-wise increasing Norfloxacin concentrations up-to 32X WT MIC. Mutants with varying resistance levels were identified with resistance being proportional to selection pressure exposed during evolution. All the evolved populations also exhibited negligible intracellular accumulation of a fluorescent dye, Ethidium bromide (EtBr) with respect to WT strain. To identify resistance determining factors, Whole Genome Sequencing (WGS) was carried out for representative colonies from all the evolved populations.

No mutation in target gene *(gyrA/B)* was identified in all the populations evolved till 4X WT MIC irrespective of the evolution profile. In case of populations evolved under different constant drug pressure, mutations in various genes were identified throughout the genome with mutations majorly in metabolic genes in low resistance mutants and mutations in energy related pathways and essential genes in high resistance mutants. Target like mutations (helicase mutation) along with ribosomal mutations appeared in populations evolved under constant drug pressure of 4X MIC. Further, the number of unique mutations identified increased depending on the selection pressure during evolution. On the contrary, populations evolved under increasing drug pressure up-to 4X MIC had comparatively lower number of mutations with mutations in and around regulatory gene mutations being dominant.

All the evolved mutations harbored a mutation in a regulatory gene, *lfr*R known as a repressor of an efflux pump LfrA in *M. smegmatis*. The mutations were spread across the gene length, with a few mutations affecting the N-terminal domain while the majority affecting the C-terminal domain. Over-expression of *lfr*R gene in the evolved populations restored accumulation levels similar to WT levels. MIC of EtBr and Norfloxacin was also restored to WT levels. Similar restoration of WT accumulation and MIC level was obtained using Efflux Pump Inhibitor (CCCP) along with EtBr indicating efflux to be the primary defense mechanism for populations evolved under 4X MIC irrespective of the evolution profile.

In case of populations evolved under increasing Norfloxacin concentrations of up-to 32X MIC, mutation in target genes (*gyr*A and *gyr*B) were identified along with mutations in other regulatory genes including *lfr*R and *crp*1. Crp1 controls gene regulation under energy limiting and hypoxic conditions in *M. smegmatis* adjusting cellular machinery to match with the energy demands of the cell. These results suggest involvement of mutations beyond efflux related mutations under conditions threatening survival.

Overall, this work demonstrates multiple adaptive trajectories for populations evolved under diverse temporal profiles with conservation of certain mutations allowing survival and assisting in further accumulation of mutations. Also, it was demonstrated that efflux mutations in general precede target gene mutations. Therefore, including EPIs in drug regimen early on during treatment process would significantly impede resistance evolution and prevent high level resistance emergence.

565e Prediction of the Formation of Co-Amorphous Systems By Means of a Machine Learning Approach

Elisabeth Fink, Michael Brunsteiner, Amrit Paudel, Sarah Zellnitz-Neugebauer

Amorphous forms and formulations of drugs have gained interest due to several favorable properties, but compared to their crystalline counterparts one disadvantage is their thermodynamic instability. Co-amorphous systems (COAMS) have been described as a promising solution to overcome this drawback and stabilize the amorphous form (Shi et al., 2019). The co-former selection is a crucial factor, however, a lack of systematic, predictive and computational methods for co-former selection has been identified (Liu et al., 2021). So far, co-former selection has mainly been done case-by-case, based on synergistic effects and/or combination therapy, previous studies, physicochemical properties or structural analysis.

This work presents a machine learning (ML) approach to predict the successful formation of new COAMS. A model based on gradient boosting methods was developed on a training dataset comprising 254 systems described in literature. A careful selection of molecular descriptors was used to obtain a numerical description of the systems. In order to increase the robustness and generalizability of our model, an ensemble learning approach was chosen in combination with cross-validation. On the training data, a correct prediction was observed in 80% of the cases and on the validation data (21 additional systems) on 76%.

Further, the model was used to predict the formation of COAMS in new, unreported systems. These consisted of two APIs as used in the field of inhalation therapy. Using our model and analysis of the training data, promising combinations of APIs, possibly exhibiting co-amorphous properties, were identified based on their likelihood of forming a COAMS. This likelihood was estimated based on the distance of the new system from the original training data and the predicted class (COAMS or not). Several of the suggested COAMS were further tested to experimentally validate the model. The developed model will expedite the experimental screening phase for new co-amorphous systems and help saving time and cost.

References:

Liu, J., Grohganz, H., Löbmann, K., Rades, T., Hempel, N.J., 2021. Co-amorphous drug formulations in numbers: Recent advances in co-amorphous drug formulations with focus on co-formability, molar ratio, preparation methods, physical stability, in vitro and in vivo performance, and new formulation strategies. Pharmaceutics 13. https://doi.org/10.3390/pharmaceutics13030389

Shi, Q., Moinuddin, S.M., Cai, T., 2019. Advances in coamorphous drug delivery systems. Acta Pharm. Sin. B 9, 19–35. https://doi.org/10.1016/j.apsb.2018.08.002

565f Engineered Anti-CD276 Mab to Deliver Payload for Cancer Treatment

Yingnan Si, Xiaoguang Liu

Targeted therapies or combined therapies are effective to treat aggressive, heterologous and recurrent cancers as compared to standard chemotherapies. This study aimed to develop an engineered anti-CD276 monoclonal antibody (mAb)-drug conjugate to treat triple-negative breast cancers (TNBCs). Specifically, CD276 surface receptor was identified as an ideal target of over 60% TNBC patients. An innovative murine anti-human mAb was developed to target the N-glycosylated extracellular domain (Leu29-Pro245) of CD276 receptor, further engineered to construct a new chimeric mAb, and conjugated with payload mertansine to construct antibody-drug conjugate (ADC). The flow cytometry analysis, confocal microscopy imaging, and In Vivo Imaging System demonstrated the capability to specifically target $CD276^{+}$ TNBC of our mAb. The IC_{50} assay showed high anti-TNBC cytotoxicity of constructed ADC in multiple cell lines. Furthermore, *in vivo* anti-tumor evaluations using immunocompetent mouse models and patient-derived xenograft (PDX) models demonstrated high tumor treatment efficacy of the anti-CD276 ADC which integrates the advantages of potency of mertansine and tumoral immunity of CD276 mAb. This study suggested that our anti-CD276 mAb-based ADC could provide a promising targeted therapy for TNBC patients in future.

565g A New Strategy for Persister Control

Sweta Roy, Zeynep Cakmak, Dacheng Ren

One major mechanism that contributes to the failure of current antibiotic treatment is the formation of persister cells, a dormant subpopulation exhibiting high-level tolerance to antibiotics. During antibiotic treatment, persisters remain arrest and transiently repopulate once the antibiotic is withdrawn, causing the failure of treatment and relapse of many infections. Recently, we reported a new strategy for persister control that leverages the dormancy characteristic of persisters and reduced drug efflux activities to increase the intracellular accumulation of the antibiotic, eravacycline. This led to effective killing of *Escherichia coli.* persisters cells (by 3 logs when treated at 100 μg/mL). This motivated us to test eravacycline on other pathogenic species such as *Pseudomonas aeruginosa*. Our results demonstrated that eravacycline is effective in killing *P. aeruginosa* PAO1 (wild-type) and PDO300 (mucoid mutant) persisters cells (99.7 ± 0.0% andamp; 99.0 ± 0.1%, respectively). We also tested the effects of eravacycline on biofilm associated persister cells given that biofilms play a major role in chronic infections. With a concentration of 100 μg/mL, eravacycline reduced the number of viable biofilm cells of PAO1 and PDO300 by 99.9 ± 0.8% and 99.6 ± 0.4%, respectively. In addition, to testing the new strategy against other pathogenic species, we utilized the results from the *E. coli* data to find new persister control agents. Here, we optimized a chemoinformatic clustering algorithm to filter a chemical compound library to a subset that are similar to the chemical properties of previously proved positive candidates that can sensitize *E. coli* persisters. From this subset, compound 1b showed potent activity against *E. coli* persister cells (killing 95.5 ± 1.7%). Collectively, these findings reveal that persister formation may cause enhanced penetration for certain antibiotics which can lead to more killing of tolerant population under normal conditions.

513a Integrated Model-Based Design and Scale-up for Peptide Chromatographic Purification

Bhoja Kandela, Bent Colin, Shekhar Viswanath

Chromatography is often an exclusive purification handle and needs to be well understood to articulate a quality control strategy. For peptide impurity rejections, the reverse-phase chromatography is the best option in comparison with other purification methods. Chromatography process modelling is well known in literature, and helps in accelerating the design, optimisation and scale-up. The objective here is to utilise "model-based approach integrated with experiments" methodology and applying it on a Lilly development asset with the objective of reduced experimentation and delivering an in-depth impurity control strategy. The methodology involves: (1) systematic experiments to determine peptide adsorption isotherm by inverse method (2) incorporating modifier effect into adsorption isotherm (3) characterisation of impurities by lumping them into pseudo-impurities (Journal of Chromatography A, 1284 (2013) 69– 79) (4) design of batch operating space using validated model (5) finally, optimising at scale with simulations for multi-column counter-current solvent gradient purification (MCSGP). The optimisation parameters selected for purification process are purity and productivity. This study helps in arriving at optimum operating space in an accelerated manner. The future work includes extending the methodology to other modalities such as oligomers and monoclonal antibodies as well as comparing various modes of continuous chromatography.

513b Model-Aided Reaction Kinetics and Rheology Study of Solid Phase Peptide Cleavage Process

Jingyao Wang, Lexie Niemoeller, Shekhar Viswanath, Yu Lu, Michael Kobierski, Bradley M. Campbell

Solid phase peptide synthesis is a widely used technique for the synthesis of medicinal peptides. In such processes, the desirable peptide sequence is built on a polystyrene resin bead via cyclic reaction sequence, *i.e.,* activation of Fmoc- protected amino acid to form activated ester, coupling of activated ester on the amine group of the peptide that attached to the resin, and then deprotection of Fmoc-protection group to expose the amine group ready for next amino acid coupling. Once synthesis is completed and the designed peptide sequence is obtained, a subsequent peptide isolation technique is frequently used where trifluoroacetic acid/dichloromethane solution mixture is applied to allow the cleavage of peptide from the resin. The peptide is then isolated via distillation, precipitation/crystallization, and drying unit operation processes. However, the peptide is subject to degradation in the acidic cleavage solution. Also, there can be significant peptide gelation issues during the cleavage reaction, where formed peptide networks dynamically increase the solution viscosity and pose significant operation risk in manufacturing. In this study, a systematic model-aided peptide kinetics and rheology study is applied to quickly identify the optimal reaction conditions using the minimum number of experiments. The reaction kinetics study was conducted with the aid of a mid-FTIR probe to track peptide and TFA solution concentrations during reaction in real-time. A detailed peptide cleavage/impurity formation kinetics model is proposed and the adsorption isotherm of TFA is considered to allow precise identification of optimal reaction conditions. Then a dynamic rheology study was applied to monitor the peptide gelation network kinetics during the soft-cleavage reaction to demonstrate acceptable solution viscosity during production process time. Combing the reaction kinetics/impurity control and gelation phase diagram, this novel approach has the potential to apply for quick rational design of any solid phase peptide isolation process.

513d Coarse-Grained Molecular Dynamics Simulation of Aggregation and Surface Adsorption of ssDNA Loaded Adeno-Associated Viral Capsids

Tibo Duran, Arani Chanda, Willow DiLuzio, Ryan Bellucci, Shivangi Naik, Bodhisattwa Chaudhuri

Purpose:

Adeno-associated virus (AAV) is a small virus with a nonenveloped icosahedral shape in the size of roughly 26 nm in diameter.) It is considered as a promising vector encapsulating a single-stranded genome (ssDNA) for *in vivo* gene delivery. However, little is known about the mechanism for the aggregation of AAV capsids during purification, their tendency to be absorbed by hydrophobic materials during the manufacturing process and storage, and the role of free ssDNA in the aggregation and adsorption processes. Accordingly, by applying a coarse-grained molecular dynamics (CG-MD) simulations, we studied the underlying mechanism of the interaction between capsids and the interaction of adsorption process between the capsid and hydrophobic surfaces, such as polyethylene (PE) and polypropylene (PP).

Methods:

The CG-MD simulations and their analyses were carried out using GROMACS package. We applied the MARTINI force-fields along with periodic boundary conditions. The crystal structure of AAV Serotype 8 (AAV8) was taken from the protein data bank entry (PDB: 6v12), and the initial 2.2kb ssDNA model was constructed using Avogadro software. The steepest descent algorithm was used in energy minimization followed by a 10 ns equilibration step in an isothermal–isochoric ensemble at ambient temperature. The simulation production runs were performed in an isothermal-isobaric ensemble using the Nosé–Hoover thermostat and the Parrinello–Rahman barostat at a pressure of 1 bar for each 20-fs time step. Computations were performed in High Performance Center Supercomputer Cluster at the University of Connecticut.

Results:

From the CG-MD simulations, the aggregation of both full and empty capsids can be captured successfully. The analysis of Root-mean-square deviation (RMSD) of all protein monomers, and root-mean-square fluctuation (RMSF) for the individual residues demonstrated that the full capsid exhibited a pronounced instability compared to the empty capsid under the same temperature and pressure conditions. However, a larger radius was captured for full capsid than that of empty capsid, which is a resultant of the increased core size of a loaded ssDNA molecule. In addition, the electrostatic potential map revealed more anionic charges on the surface of the full capsid than that of empty capsid due to the encapsulated negatively charged ssDNA in the full capsid. Moreover, the behavior of the full capsid adsorbed onto the hydrophobic surface was captured successfully. It was noted that the capsid was rotated, driven toward, and attached to the surface with stronger interactions to the PP surface than that of the PE surface, which was assessed by relative shape anisotropy.

Conclusions:

A better understanding of capsid aggregation associated with multiple asparagine and glutamine residues was established by CG-MD simulations. Additionally, multiple hydrophobic residues, such as leucine, isoleucine and phenylalanine, were identified to be related to the surface adsorption process. The stability and electrostatic charges of the capsids and the free ssDNA molecule were evaluated using this developed computational model. Our simulations provide general dynamical information about the structure of the ssDNA and capsids on microsecond timescale, revealing behavior that is impossible to be obtained from experimental studies. The CG-MD simulations are demonstrated to be a powerful tool in the simulation of macromolecular complexes and are capable of guiding the interpretation, explanation, and direction of experiments.

513e Leveraging Molecular Modeling in the Selection of Protein-Drug Candidates for Late-Stage Developability

Marco Blanco, Francis Insaidoo, Jainik Panchal, Veronica Juan, Marc Bailly

Understanding and controlling the stability of protein solutions is a particularly relevant and current issue in the biopharmaceutical pipeline. Proteins are marginally stable and are susceptible to undergo multiple degradation routes such as aggregation, phase separation and oxidation, which poses a significant challenge for the development and manufacturing of biotherapeutics. Traditionally, these instabilities are mitigated by minimizing protein interactions via changes in processing and formulation conditions (e.g., adjusting pH, addition of excipients). While this strategy has been proven successful for advancing biotherapeutics to clinical trials and subsequently to help patients, it generally requires large amount of resources (both time and material) to identify suitable processing and phase-appropriate solution conditions. In this regard, there is a need for developing complementary approaches that reduce the time and cost for the advancement of biologics and accelerate bringing drug products to patients. This talk explores one of such approaches that relies on the use of computational tools for selecting biologic drug candidates with minimized risks for formulation and process development. Specifically, this work aims to guide the early large-molecule development to modulate not only their potency and efficacy but also their interactions to enhance their physical and chemical stability. This talk illustrates the implementation of different fundamental and statistical approaches (e.g., molecular modeling and machine learning) to identify the presence of interacting regions or 'patches' on the protein surface, which are related to instability problems in biotherapeutics. Moreover, the results show different strategies based on statistical modeling and atomistic and coarse-grained simulations to identify key residues in the protein surface, which yield the highest impact in the distribution of protein surface patches. These results are contrasted against experimental biophysical and biochemical characterization of the thermodynamic stability and aggregation propensity of multiple point-mutations of a model monoclonal antibody.

513f Design of Experiments Compared to Artificial Neural Network Approach – How Can We Use Them to Understand Spray Drying of Biologics?

Daniela Fiedler, Elisabeth Fink, Isabella Aigner, Eva Roblegg, Johannes G. Khinast

The therapeutic potential of biologics with the advances made in biotechnology leads to a wide variety of biologicals on the pharmaceutical market and in the pipeline. Most of these valuable and complex entities are currently formulated as a liquid final dosage form. However, the transformation into a solid-state offers the benefits of longer storage times and higher storage temperatures. The gold standard process for this liquid to solid transformation is lyophilization. Lyophilization offers advantages, such as low temperatures during drying, large knowledge available in the literature, and filling and drying in the final containment. However, lyophilization is a batch process, with flexible processing and up-scaling being major issues, combined with the risk of activity loss for the biologic itself through, e.g., pH shift, cold denaturation, ice-liquid interfacial denaturation, etc. (Sharma et al., 2021). One of the most promising alternatives is spray drying. In this process, the feed solution is atomized into small droplets. Then, the solvent is evaporated gently through mass transfer from the drop to the surrounding drying gas stream. Finally, the dried powder can be collected after a separation step (e.g., cyclone). This drying process can be performed continuously, resulting in dried active biologics of therapeutic interest (e.g., Interleukins (Fiedler et al., 2021)) with more flexibility in throughput and upscaling.

Process development of a spray drying process requires several grams (to kilograms) of the active pharmaceutical ingredient (i.e., biologic). However, on the one hand, their prices can be in the range of several hundred dollars per milligram, and on the other hand, the production of biologics may be in the range of several kilograms per year (Pennington et al., 2021). Hence, developing a spray drying process can be unfeasible due to the lack of available material and high development costs. Thus, a cheap surrogate substance was investigated in our approach to overcome this hurdle. Clearly, the most critical part of such an approach is finding a suitable surrogate substance with a similar drying performance as the protein. However, identifying relevant material attributes for a drying process is challenging due the complex phenomena observed, including skin formation on the droplet and water diffusion in the particle and through the skin, which significantly influences drying performance. Therefore, we started with Lactose as surrogate substance as it widely used in spray drying for process validation runs during equipment setup and testing. Process data from such tests are often available.

Hence, a first Design of Experiments (DoE, 2^4 full-factorial) was performed with Lactose. The data gained from this DoE reflects the drying kinetics and evaporation capacity of the spray dryer (Buchi B-290) at different process settings. Eventually, only three spray drying experiments were performed with the protein (human serum albumin, HSA) to transfer the process dynamics including the protein. Therefore, we used the DoE data to develop an Artificial Intelligence Approach (using artificial neural networks ANN), which we compared to and combined with the classic DoE approach.

From the lactose DoE the derived models primarily depend on the number of investigated factors and the factor combinations and the number of experiments performed, resulting in linear and quadratic models (as well as considering the interaction of factors). The capability of the resulting models to predict the responses is limited, especially when the underlying physical correlations are complex and interdependent. Moreover, the Critical Quality Attributes (e.g., residual moisture, particle size distribution, yield, etc.) were investigated and examined concerning their ability to predicted reliably. Additionally, the HSA was analyzed concerning its secondary structure to derive an understanding of the limits of proteins processed at mild temperature spray drying conditions (i.e., outlet temperature less than 100°C). When it comes to the drying of biological entities, their influence on the drying kinetics (e.g., the binding capacity of water, diffusion in the liquid drop) is not well understood. Therefore, we investigated the influence of protein concentration on the evaporation kinetics by means of thermogravimetric analysis.

This talk will highlight the intrinsic opportunities of an ANN approach compared to a DoE approach for developing spray drying processes for biologics. With this, it is possible to fully exploit the advantages of a continuous drying process. Furthermore, we show a method that allows the determination of the drying kinetics of different protein concentrations at mild temperatures, including first results for HSA. This method can also be used to test surrogate substances and create a database for which will in the future allow even better fitting of the developed models. Initial experimental results highlight the potential and benefits of a combinatory planning approach, by requiring fewer experiments, and thus significantly reducing the material and time during process development.

Literature:

Fiedler, D., Hartl, S., Gerlza, T., Trojacher, C., Kungl, A., Khinast, J., Roblegg, E., 2021. Comparing freeze drying and spray drying of interleukins using model protein CXCL8 and its variants. Eur. J. Pharm. Biopharm. 168, 152–165. https://doi.org/10.1016/j.ejpb.2021.08.006

Pennington, M.W., Zell, B., Bai, C.J., 2021. Commercial manufacturing of current good manufacturing practice peptides spanning the gamut from neoantigen to commercial large-scale products. Med. Drug Discov. 9, 100071. https://doi.org/10.1016/j.medidd.2020.100071

Sharma, A., Khamar, D., Cullen, S., Hayden, A., Hughes, H., 2021. Innovative Drying Technologies for Biopharmaceuticals. Int. J. Pharm. 609, 121115. https://doi.org/10.1016/j.ijpharm.2021.121115

513g Investigating Fundamental Motions to Rapidly Predict Lyophilized Protein Stability

Kelly Badilla, Andreas Bommarius, Marcus T Cicerone

While it is common knowledge that lyophilization serves as a technique to stabilize protein-based drugs, the mechanism by which this occurs is not yet entirely known. Thermodynamics and reaction kinetics both play roles in the destabilization and eventual degradation of the proteins contained within the glass system, but the precursor to both degradation pathways is fundamental molecular-scale motion. Juxtaposed to the long-time studies that are currently used to quantify the stabilizing capabilities of their respective excipients, I will present how probing the fundamental motions can rapidly provide insight about protein stability for lyophilized systems.

It has been shown that there is a direct correlation between the degradation rate of proteins stored in the glass phase and molecular mobility of the systems in which the proteins are embedded[1]; this correlation holds true for a variety of proteins stored in multiple excipients and at multiple temperatures. This work was conducted using neutron scattering which is a technique less accessible to most researchers. However, alternative methods exist that allow for acquisition of similar data. The first method discussed in the present work is measurement of the optical Kerr effect (OKE) from which molecular reorientation times collected under the stress of an ultrashort laser pulse can be related to molecular mobility. It has been shown that measurement of the OKE provides signatures of the fundamental motions[2], similar to observations from neutron scattering. The second method discussed in the present work is low-frequency Raman spectroscopy from which the energy difference of inelastically scattered light can be related to molecular mobility. Both these techniques are expected to provide insight on the stabilizing power of excipients in a much shorter window than current long-time studies because they measure the most fundamental motions of the system, which are sub-steps of the processes that cause protein degradation. These techniques can serve to drastically decrease the time spent optimizing formulations.

1. Cicerone et al., *Soft Matter,* **8** (2012)
2. Bender et al., *Soft Matter,* **16** (2020)

32a Tailored Composition of Lipid-Based Excipients for Improved Functionality in Advanced Pharmaceutical Manufacturing Processes

Sharareh Salar-Behzadi, Carolina Corzo, Moaaz Abdelhamid, MSc, Martin Spörk, Ana Belén Ocampo, Mira Maisriemler, Dirk Lochmann, Sebastian Reyer, Tanja Freichel

Background and objective

The key requirement for the development of innovative and robust dosage forms is the application of excipients with desired functionality. In this context, there is a growing interest in using lipid-based excipients due to their low toxicity, biocompatibility and biodegradability. The challenge is however, their unstable solid state and the high degree of crystallinity, affecting the stable performance of their products and limiting their application for certain manufacturing processes, respectively.

In previous works, we introduced polyglycerol fatty acid esters (PGFAs) available as Witepsol® PMFs as advanced lipid-based excipients with ideal properties, such as being monophasic systems with stable polymorphic form (1, 2). In the current study, the effect of minor fractions in the composition of PGFAs has been investigated on their solid state and mechanical properties. The results have been used for tailoring the composition with defined solid state to reach ideal properties for challenging applications, such as melt extrusion and filament-based 3D printing.

Materials and methods

PGFAs are hydroxyethers of glycerol, fully or partially esterified with saturated fatty acids. As same as most other lipid-based excipients, their synthesis yields a composition of similar molecules, having defined PGFA as the main component and minor fractions of polyglycerol mono esters and free polyglycerols (PG). In this study, PG6C16p; composed of six glycerol moieties partially esterified with palmitic acid was used. Two different batches of PG6C16p with different concentrations of minor fractions were screened. The batch with higher concentrations of PG monoester and free PG was selected for further screening. A variety of free PG(n) molecules, composed of two, three, four and six glycerol moieties (n=2 to 6) were provided and systematically added to the selected PG6C16p batch, in three different concentrations; $2.6\%_{w/w}$, $8.6\%_{w/w}$ and $14.6\%_{w/w}$.

Melt casted samples were provided from the PG6C16p batch (blank) and each mixture. The solid state of melt casted films in terms of thermal behavior, polymorphism, and lamellar configuration was analyzed via DSC and x-ray diffraction after preparation (T0) and after storage under long term (25°C, 65% relative humidity) and accelerated (40°C, 70% r.h.) conditions. Processability of mixtures in melt extrusion, the solid state of filaments and their quality in terms of flexibility, filament diameter and diameter distribution were investigated and compared with each other.

Results and discussion

Melt casted samples

<u>Differential Scanning Calorimetry (DSC)</u>: The DSC data showed the similar thermal behavior of PG6C16p batch (blank) to its mixtures with different free PG. All samples displayed two endothermic peaks at T0, which were dropped to one phase after storage under both conditions. The cooling cycle also revealed two crystalline phases. The area under the curve of the cooling cycle was found to decrease with the addition of free PG, depicting less crystallinity compared to the blank (PG6C16p

batch).

X-ray diffraction: The x-ray analysis in wide angle region showed a stable α-form (no polymorphism) for the blank as well as for its mixtures with free PGs. This phenomenon is typical for all PGFA molecules and described elsewhere (1).

The existence of two crystalline phases in PG6C16p batch (blank) and its mixtures with free PG was confirmed with the x-ray data in small angle region, showing two coexisting phases (hexagonal and lamellar arrangements) with two categories of d-spacing. Phase transitions toward a single lamellar phase was observed after storage, which can be explained due to existence of free hydroxyl groups with high level of polarity, triggering the system towards the phase transition. Further studies on the kinetics of phase transition showed a tempering step at 40 °C for approximately 20 minutes is sufficient for the complete phase transition to a stable lamellar phase.

Elongation of the lamellar length of PG6C16p (d-spacing) was noticed with the addition of free PGs in the following order: free PG6<PG4<PG3<PG2. Addition of free PG2 displayed the maximum elongation, whereas, free PG6 displayed the minimum elongation. This can be explained with the smaller molecular size of free PG2 compared to free PG6, providing an easier submerging of free PG2 between the lamella layers of the main component (PG6C16p) as well as the molecular similarity of free PG6 to PG6C16p. The molecular similarity was also responsible for the negligible crystal growth of main component in its mixture with free PG6 after storage under both long term and accelerated conditions. In contrary, samples of PG6C16p with additional free PG2 showed a significant crystal growth.

Melt extrusion of samples and filament quality

PG6C16p batch was processable via melt extrusion. However, upon increasing the percentage of free PG, or the number of glycerol units or both, improved processing and improved quality of the produced filaments were observed (e.g. ease of process adjustment, homogeneity and flexibility/spoolability of filaments). This can be explained by the reduced crystallinity of the main component in front of free PGs (observed in the DSC thermograms of melt casted films, explained above) and the fact that the increase of free PG in the system is associated with an increase in available sites for hydrogen bonding, which may contribute to the enhanced filament flexibility.

The WAXS analysis of filaments after manufacturing and after storage confirmed the stable α-form also in filaments. The Phase transitions toward a single stable lamellar phase also took place during the process under the effect of time, temperature, and shear.

Conclusion

A mechanistic knowledge of the composition of lipid-based excipients is essential to understand their functionality. The composition of a candidate molecule (PG6C16p) belonging to the group of PGFA excipients was successfully tailored through the fine-tuning of its minor fractions. The molecular similarity between the main component and minor fractions was essential for providing stable solid state. The smart balance between the components resulted in reduced degree of crystallinity in the main component. Adjusted degree of polarity in the system provided extra potential sites for hydrogen bonding. These two behaviors were the key parameters for improving the processability via melt extrusion. The resulted filaments showed advanced properties for 3D printing. This work is an important step towards the extending the functionality of lipid-based excipients for advanced pharmaceutical manufacturing processes.

Acknowledgement

The Austrian Research Promotion Agency (FFG)

References

1. Corzo C. et al., (2020), European Journal of Pharmaceutics and Biopharmaceutics, 148, 134–147. https://doi.org/10.1016/j.ejpb.2020.01.012
2. Salar-Behzadi S., et al., Structure - function analysis of lipids for the development of advanced excipients for pharmaceutical manufacturing, AIChE Annual Meeting 2021

Keywords

Pharmaceuticals (Chemicals andamp; Materials), Lipid-based excipients, Structure-Function of excipients; Pharmaceutical Engineering andamp; Drug Delivery, Pharmaceutical Development

32b Formulation and Scale-up of Delamanid Nanoparticles for Oral Tuberculosis Treatment

Nicholas Caggiano, Joanna Georgiou, Madeleine Armstrong, Rodney Priestley, Robert K. Prud'homme

Delamanid is a small molecule anti-tuberculosis therapeutic used to treat multi-drug resistant tuberculosis (MDR-TB), a strain of the disease associated with higher mortality rates and the need for harsher therapies. However, the bioavailability of delamanid is limited by its hydrophobicity and crystallinity. Nanoparticle encapsulation is an attractive route to increase the bioavailability of delamanid by increasing the specific surface area available for drug dissolution and by forming an amorphous drug core.

Initial investigation of nanoprecipitation as a formulation route revealed significant incompatibility between the solid drug core and commonly employed nanoparticle stabilizers. To address this, emulsification was investigated as a route to form stable drug-loaded nanoparticles while avoiding the *in-situ* formation of a solid drug core, as the delamanid remained solubilized in a liquid core during emulsification. Inexpensive, naturally derived emulsifiers, including lecithin and functionalized cellulosics, were employed as stabilizers. Tensiometry measurements were conducted to quantify the effect of each stabilizer on the oil/water interfacial tension. Initial formulation screening was conducted at batch scale using probe tip sonication, and scale-up studies were conducted using high pressure homogenization. The size stability of the resulting nanoparticles was characterized over time using dynamic light scattering. Depending on the stabilizer used, emulsification produced stable particles with average diameters ranging from 100-600 nm. Subsequent spray drying and redispersion studies were conducted to produce a dried powder. Combinations of lecithin and cellulose stabilizers were found to be beneficial in producing smaller particles (~250 nm) which were also robust enough for spray drying without significant aggregation. An *in vitro* dissolution assay revealed that the dried emulsion formulations produced enhanced dissolution kinetics compared to bulk crystalline delamanid.

32c Preparation and Characterization of Modafinil Nanofibers By Electrospinning

Indumathi Sathisaran, Jean-Christophe Monbaliu, Cornelis Vlaar, Jorge Duconge, Vilmalí López-Mejías, Torsten Stelzer

Modafinil is an anti-narcoleptic drug used to treat excessive daytime sleepiness associated with sleep apnea and shift-work disorder.[1] It has been approved by the Food and Drug Administration (FDA) and is being marketed in the form of tablets with a trade name, 'Provigil'.[1] Modafinil tablet forms suffer from disadvantages of susceptibility of reduced solid-state stability and drug loading. As nanofibers possess exceptional mechanical, physical and chemical properties, these formulations have gained tremendous attention in the scientific community.[2] For the first time, in this work, we report the formulation of polymeric nanofibers loaded with modafinil at dosage recommended by FDA (100 mg) via electrospinning. The modafinil-loaded polymer nanofibers were prepared by electrospinning polymer solutions with suspended modafinil crystals at a voltage of 12 kV and a flow rate of 5 mL/h. Polymers used were Poly Vinyl Pyrrolidone (PVP), Poly Vinyl Alcohol (PVA), Poly Ethylene Glycol (PEG), and Poly L-Lactide (PLLA). The obtained electrospun nanofiber formulations were characterized for their morphological, vibrational, structural, and thermal properties by Scanning Electron Microscopy (SEM), Raman spectroscopy, Powder X-Ray Diffraction (PXRD), Differential Scanning Calorimetric (DSC) and Thermogravimetric Analysis (TGA). The modafinil loading in the nanofibers and the dissolution of the electrospun modafinil nanofibres were determined using UV-Vis spectrophotometer at 220 nm.[3] The morphological analysis revealed the encapsulation of circular-shaped modafinil particles within the nanofibers further evidenced by the broader stretching of the Raman spectral region around the wave numbers ranging between 1425 cm^{-1} – 1225 cm^{-1}, 1050 cm^{-1}-1000 cm^{-1} and 650 cm^{-1} – 620 cm^{-1} representative for C-H stretching, presence of a polystyrene peak and a S=O stretching in modafinil. The PXRD analysis confirmed the presence of modafinil in the nanofibers and the crystalline nature of the formulation. The melting point of the modafinil-loaded nanofibers was found to be higher than the melting point of commercial modafinil indicating that the modafinil-nanofiber formulation possesses a good thermal stability characteristic than the commercial modafinil. The unoptimized loading of modafinil into the nanofibers was determined to be 65%. The modafinil-loaded nanofibers illustrate that electrospinning can serve as a less laborious and an easily adoptable continuous method for the formulation of modafinil according to the desired dosage needs. Consequently, this study might be used as a rationale for necessary further analytical and regulatory studies.

Keywords: Modafinil, electrospinning, loading.

References

(1) Minzenberg, M. J.; Carter, C. S. Modafinil: A Review of Neurochemical Actions and Effects on Cognition, *Neuropsychopharmacology* 2008, 33, 1477-1502.

(2) Morie, A.; Garg, T.; Goyal, A. K.; Rath, G. Nanofibers as novel drug carrier, *Artificial Cells, Nanomedicine, and Biotechnology* 2014, 44(1), 135-143.

(3) Adibika, K.; Selselehjonban, S.; Emami, S.; Osouli-Bostanabad, K.; Barzegar-Jalali, M. Electrosprayed polymeric nanobeads and nanofibers of modafinil: preparation, characterization, and drug release studies, Bioimpacts 2019, 9(3), 179-188.

32d Impact of Material Attributes and Process Parameters on Low Dose Powder Filling Using Drum Filler System

Tanu Mehta, Zhanjie Liu, Bruhal Shah, Justin P. Lacombe, Bodhisattwa Chaudhuri

Purpose:

Low dose powder filling forms an important component of dry powder inhaler (DPI) manufacturing. There are multiple methods exist for low powder filling such as tamp filling, dosator filling, vacuum dosator filling and vacuum drum filling but powder filling in milligram fill weight is still a challenge. As most of the pharmaceutical powders are poor flowing and show unpredictable flow behavior, this filling process becomes even more difficult. In this study, we have evaluated a vacuum drum filling system by Harro Hofliger to understand the impact of material attributes (density, particle size distribution and flowability) and process parameters (vacuum pressure, drum bore size, and stirrer type) on in-process controls (fill weight and %RSD) for capsule filling process.

Methods:

Pharmaceutical powders with different material attributes were selected and filled in size 3 capsules using Harro Hofliger's DrumLab (table 1). Lactohale 300/LH300, Inhalac 230 and Inhalac 120 are poor, fair and good flowing powders respectively. We have also evaluated an in-house formulation A. unlike other selected materials, proprietary formulation A was highly cohesive and low-density (<100g/l) demonstrating very poor flow properties. As shown Table 1: Material attributes of selected powders

Powders were filled at different vacuum pressures using drum bore size 7.5mm^3 and 30mm^3, and net fill weight of each capsule was recorded to study fill weight variation (%RSD) and product density (mg/mm^3). Additionally, the impact of stirrer geometry on fill weight and %RSD was also studied using two stirrer geometries (standard stirrer and asymmetric stirrer)

Results:

Study demonstrated that fill weight was a function of vacuum pressure and linear relationship was observed between fill weight and vacuum pressure (table 2) for all materials except Inhalac 120. However, higher %RSD was observed with LH300 and formulation A. Additionally, asymmetric stirrer led to lower %RSD and higher dosing compared to standard stirrer for LH300 and formulation A which could be attributed to its paddle shaped geometry.

Conclusions:

Drum filling is an efficient method for low dose filling and can be easily scaled from laboratory to pilot scale. However, it is important to understand the material attributes to determine process parameters such as vacuum pressure, drum bore size and type of stirrer to optimize overall filling process.

Table 1: Material attributes of selected powders				
Material	**Tap Density(g/l)**	**Bulk density(g/l)**	**D50 (µm)**	**Compressibility Index (%)**
Lactohale/LH300	520	260	<5	>25
Inhalac 230	850	700	70-110	18
Inhalac 120	830	720	110-155	13
Formulation A	-	90.44	-	43.4

Table 2: Fill weight and %RSD for LH300 using 30mm3 drum					
Material	**Vacuum (mbar)**	**Standard Stirrer**		**Asymmetric Stirrer**	
		Fill weight (mg)	%RSD	Fill weight (mg)	%RSD
LH300	200	11.031	7.353	11.261	7.370
	400	13.531	7.499	14.453	3.774
	600	14.846	5.321	15.222	3.442

Results: Lactohale 300 (Fill weight (mg) vs. Vacuum (mbar))

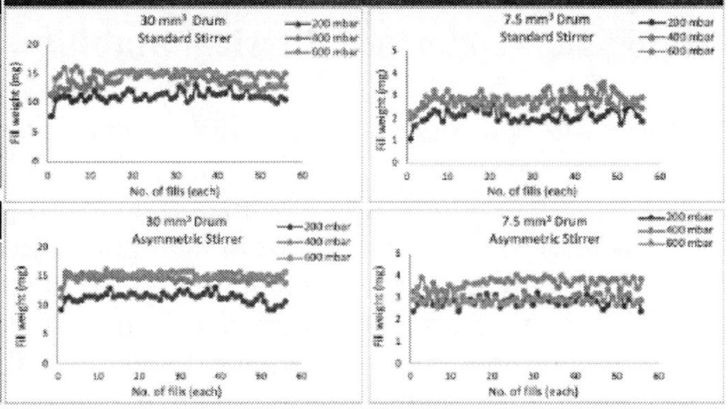

32e Sepineo P600- Based Topical Semisolid Formulation- Relationship between Rheological Properties, Physical Stability, and Drug Release

Abu Zayed Md Badruddoza, Thean Yeoh, Dana Gates, Jaymin Shah

The emulsion-based topical semisolid dosage forms present a high degree of complexity in their microstructures which is evident from their compositions made up of two or more immiscible phases as well as from mixing and interactions among multifunctional excipients and active substance. Rheological parameters are very crucial not only in determining the physical stability of the formulation during product development and the shelf life but also in influencing the release and skin permeation of drugs. In this present study, we develop Sepineo P600 based gel and emulsion gel (emulgel) topical formulations and evaluate the relationship between the rheological properties, physical instability due to phase separation, the *in vitro* release of drugs. We find that formulation parameters such as pH, nature of APIs and their concentration, concentration of dispersed phase and polymers can significantly affect the rheology as well as the physical stability and drug release. We also demonstrate rapid assessment and prediction of the long-term physical stability and shelf life of topical semisolid products using a dispersion analytical tool namely 'LUMiSizer'. This study provides valuable insights into rheological behaviors of Sepineo P600 gel and emulgel which can be used to construct a framework for designing optimal topical drug delivery system.

High Throughput Sterile Filtration of Highly Viscous Pharmaceutical Formulations

Mohannad J. Kadhum, Jack E. Kochevar, Ben Koch, Ryan K. Swanson, Steve W. Laninga*
Lifecore Biomedical, MN, USA
**Corresponding author: mohannad.kadhum@lifecore.com*

Sterile filtration using 0.2um membranes is a widely used method to eliminate the microbial load in a wide range of pharmaceutical formulations. However, sterile filtration of highly viscous solutions is challenging due to low throughput and potential concentration changes stemming from solution viscosity, retention or adsorption. Due to the inherent difficulty in filtration of relatively large molecular weight polymers or highly concentrated biologics, other sterilization methods such as e-beam, gamma radiation, dry heat, autoclave and others are usually implemented, if applicable. These methods may result in undesirable product attribute changes. Sterile filtration is usually restricted by manufacturer operating ranges of single use cartridges and capsules. The limitation of using such filters are usually low operating pressure up to 60 to 90 psi in addition to significant holdup or concentration changes by size exclusion or adsorption.

In this work, high pressure sterile filtration (HPSF) is presented as a method to increase yield without compromise on patient safety or product quality. A validated HPSF system that adheres to PDA and ASTM sterile filtration guidance was utilized (1, 2). This system has been used to filter sterilize sodium hyaluronate (NaHy) solutions used as ophthalmic viscoelastic devices (OVD), which are used in cataract surgery (nominal viscosities of 50,000 cps). Recently, the application of HPSF has expanded to other highly viscous solutions. The HPSF system runs up to a 900 psi for filtration of highly viscous formulations of NaHy, sodium carboxymethyl cellulose(Na-CMC), and other widely used excipients.

Formulations of NaHy with molecular weights ranging from 900 up to 2,000 kDa with concentrations as high as 20g/L were examined. When comparing the filtration yield of a 10g/L NaHy solution with 900kDa molecular weight on a capsule filter at the maximum operating pressure (60 psi) set by the manufacturer to a HPSF with nearly equivalent filtration surface area of 200cm^2, the flux on the high pressure filter was about three orders of magnitude faster for the same solution. Furthermore, when comparing the NaHy concentration of filtrate to the feed using the HPSF to off the shelf filter cartridges, the HPSF maintained the concentration and viscosity of the feed solution while a single cartridge caused 7% drop in concentration and up to 15% when redundant filtration was used. This clearly demonstrate the benefits of HPSF when filtering NaHy formulations.

Experiments conducted using solutions of 10g/L of NaHy of 900kDa molecular weight were filtered at pressures ranging from 45 up to 600 psi. Figure 1a shows that while filtration at pressures up to 95 psi resulted in low throughputs, high pressure continued to filter without plugging. Figure 1b shows that the maximum throughput at 600 psi was about 2 orders of magnitude higher than the throughput at 45 psi.

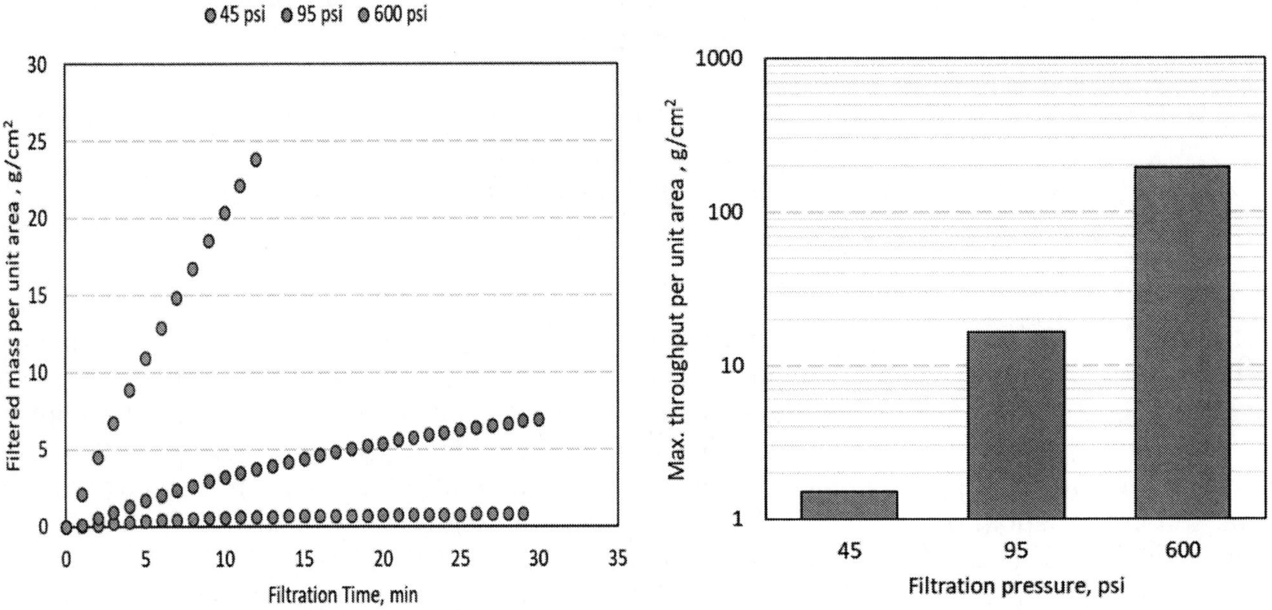

Figure 1 Sterile Filtration Throughput of NaHy solution of 900kDa and 10g/L concentration at varying filtration pressures a) filtered mass per unit filter area vs time of filtration b) comparison of the maximum throughput achieved at three filtration pressures.

A 20g/L NaHy solution with a viscosity of about 300,000 cp was successfully filtered at 800 psi, but was mainly unfilterable at low pressure of 45 and 95 psi as shown in Figure 2a. The viscosity of both unfiltered solution and filtrate of 800 psi filtration was the same as shown in Figure 2b indicating no concentration change or retention by the membrane.

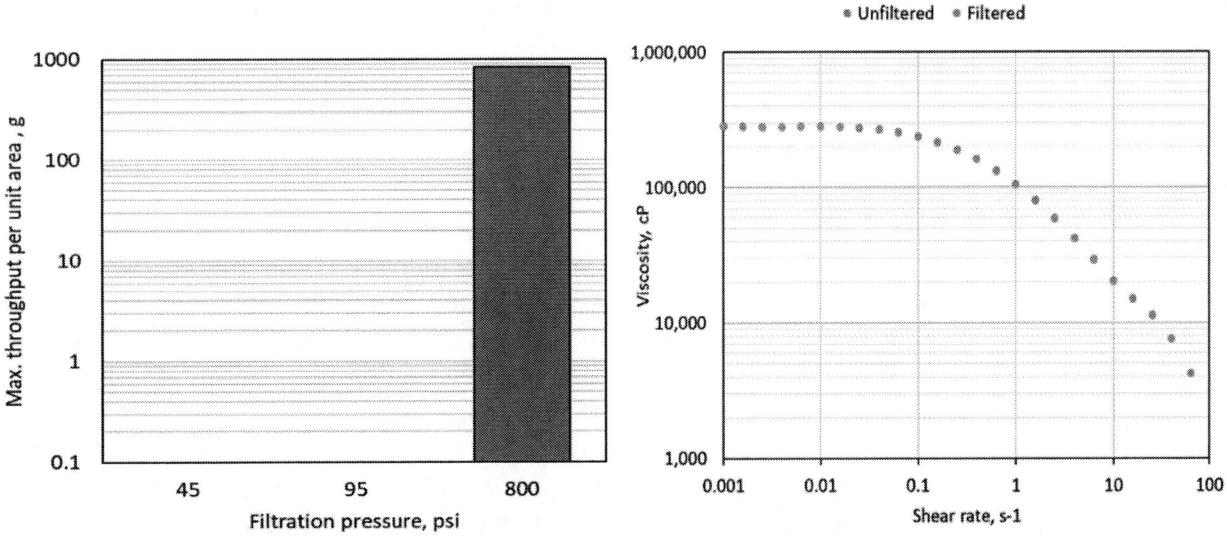

Figure 2 Sterile Filtration Throughput of NaHy solution of 900kDa and 20g/L concentration at varying filtration pressure a) filtered mass at range at tested pressures b) flow curves of unfiltered sample and 800 psi filtrate.

Additional work using 2,000 kDa NaHy solution of 10g/L also demonstrated higher throughput and no change in viscosity of both filtered and unfiltered solutions (Figure 3). The maximum throughput is shown to be 100-200 times higher at 800 psi in comparison to 45 psi filtration.

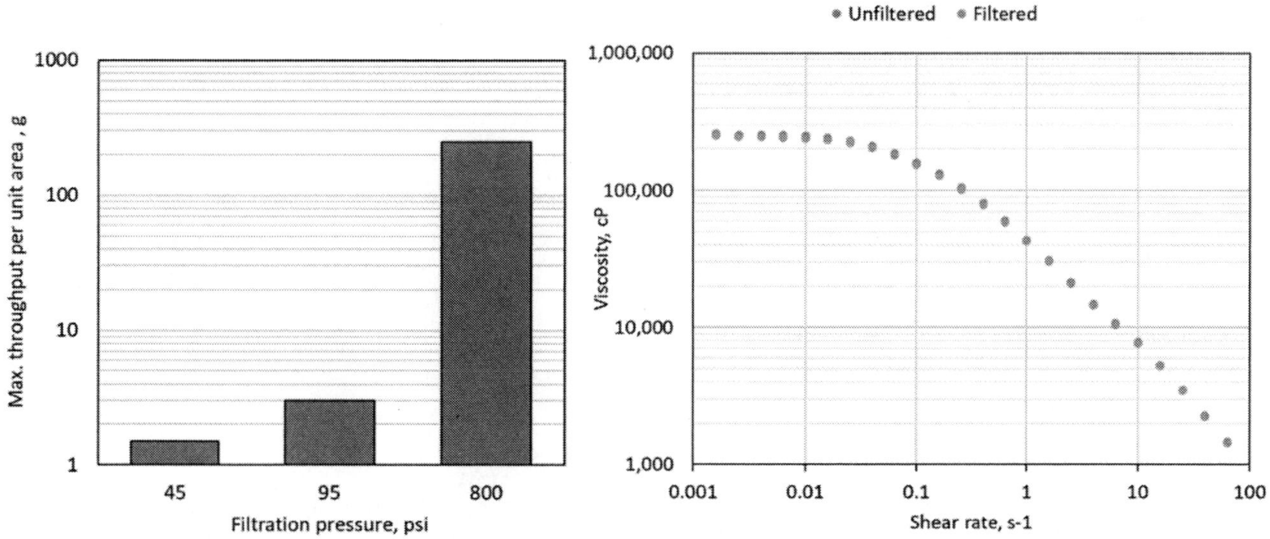

Figure 3 Sterile Filtration Throughput of NaHy solution of 2000kDa and 10g/L concentration at varying filtration pressure a) maximum throughput achieved b) flow curves of unfiltered sample and 800 psi filtrate.

Generally, the data above demonstrates that high pressure filtration is successful in filtering samples without a change in concentration. In comparison at low pressure, filtering a 10g/L solution of 900kDa molecular weight NaHy using an off the shelf cartridge with ≈2700 cm^2 effective surface area resulted in significant retention of NaHy. Figure 4 below shows that a drop of about 25% in concentration was encountered when measured using a loss on drying(LOD) method using a halogen moisture analyzer. The concentration of samples collected during the entire filtration did not reach feed concentration suggesting constant retention of NaHy which can significantly complicate manufacturing processes. Retention of polymer and concentration change complicate design of pharmaceuticals formulation at low pressure especially when redundant filtrations (two filters in series) are used.

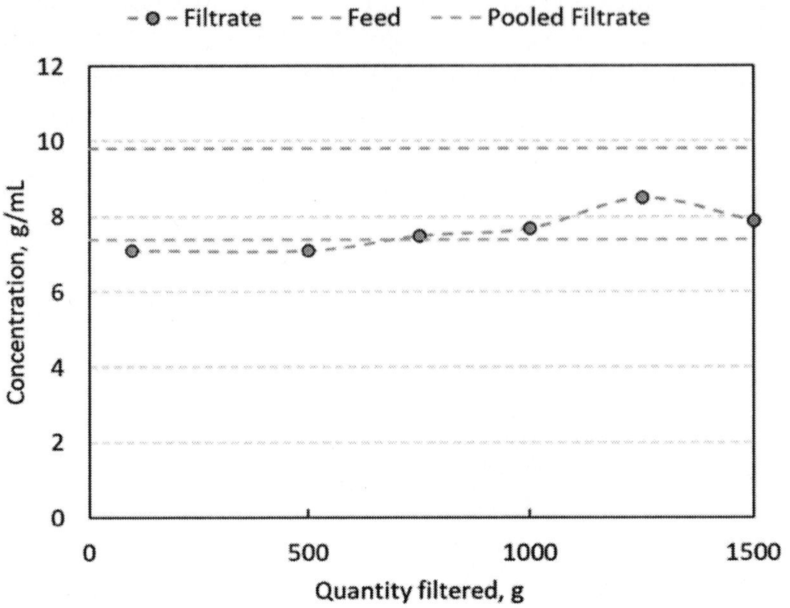

Figure 4 Concentration change between filtered and unfiltered solution at low pressure of 40 psi using 2700cm^2 PES filter cartridge.

Formulations of Na-CMC with molecular weights of 250kDa and 700 kDa with 0.7-1.2 degree of substitution of carboxymethyl and up to 20g/L in concentration were also examined. A 2% CMC formulation has on average a one order of magnitude higher throughput utilizing the HPSF system operated at 600 psi in comparison to filtration at 95 psi utilizing equivalent filtration surface areas as shown in Figure 5.

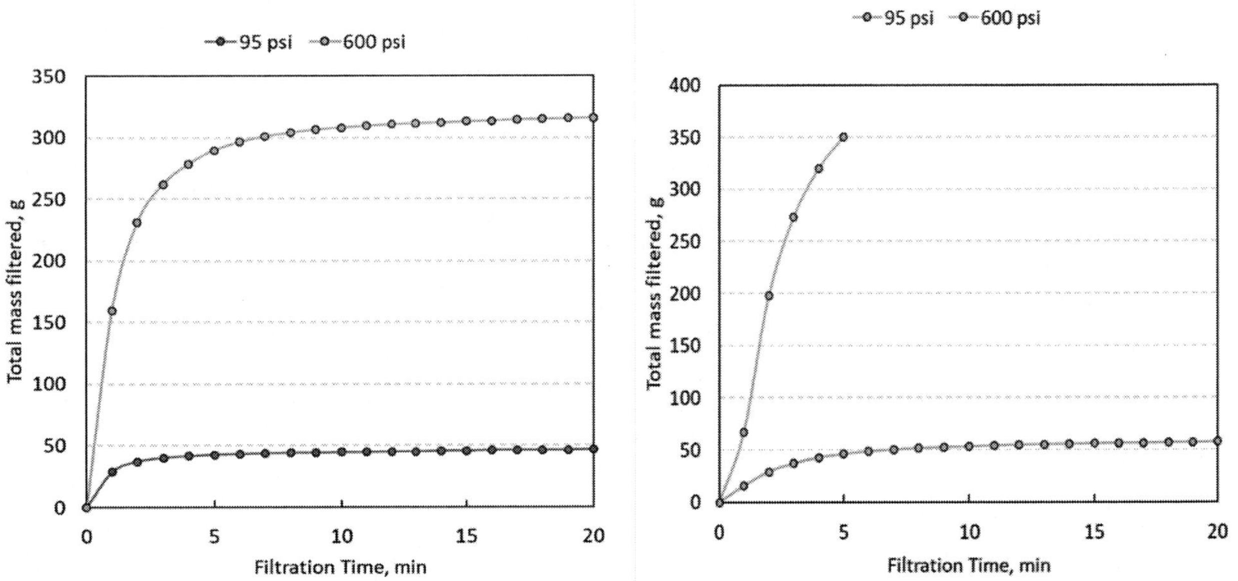

Figure 5 Filtration of NaCMC solution at range of pressures a) 5g/L NaCMC with 0.7 degree substitution b) 20g/L with 1.2 degree substitution.

In summary, sterile filtration of highly viscous pharmaceutical formulations using HPSF system can have clear advantages in terms of higher throughput, higher flux, and reproducibility in terms of concentration, and lower material holdup which can be beneficial when handling costly formulations. High pressure sterile filtration can fast track development effort and accelerate bringing early phase formulations into the clinic.

References

1. Sterilizing Filtration of Liquids, PDA Journal of Pharmaceutical Science and Technology, Technical Report No. 26 (2008).
2. Standard Test Method for Determining Bacterial Retention of Membrane Filters Utilized for Liquid Filtration, ASTM International, F838-20(2020).

32g Parameters for Controlling Liposome Particle Size By Extrusion

Maoqi Feng, Songqing Lu

Liposome is a proven effective drug delivery tool including Covid-19 vaccine. The liposome particle size is very important for the effectiveness of drug delivery. In this paper, we present the parameters which control the liposome particle size from our recent test results of formulating mRNA and proteins with liposome. The parameters include extrusion temperature, filter membrane pore size, pH, static charge, molecular weight of the mPEG in the formulation, etc. The size of the liposome particles in a cuvette was measured using Malvern Nano Ultra dynamic light scattering instrument at 25°C. The results are quite useful to drug formulation and vaccine development.

606a The Importance of Solid-State Modeling in Drug Development

Luca Iuzzolino

The properties of organic solids, including physiochemical stability, solubility, and mechanical behavior, are drastically affected by the way in which molecules arrange themselves. One fundamental component of small-molecule drug development is to determine and characterize the crystalline forms (or polymorphs) of the active pharmaceutical ingredient (API), as well as its solvated forms, and choose one that possesses an optimal set of properties. When the pure API or its solvates cannot form crystal structures with suitable properties (e.g. sufficient solubility), alternative solid forms, like salts or cocrystals, can be explored to strike an optimal balance. Thorough experimental form screens are costly, laborious and yet not guaranteed to find all the important solid phases. Solid-state modeling can be of tremendous value to drug development, as it can help right-size the experimental effort, assess the presence or absence of risks associated with the chosen form and direct experimental work towards the most relevant threats and opportunities. In this presentation, we show how solid-state modeling, including crystal structure prediction and virtual cocrystal screening, can support drug development projects, and how its integration with experimental efforts has made it a vital component in the solid form selection process.

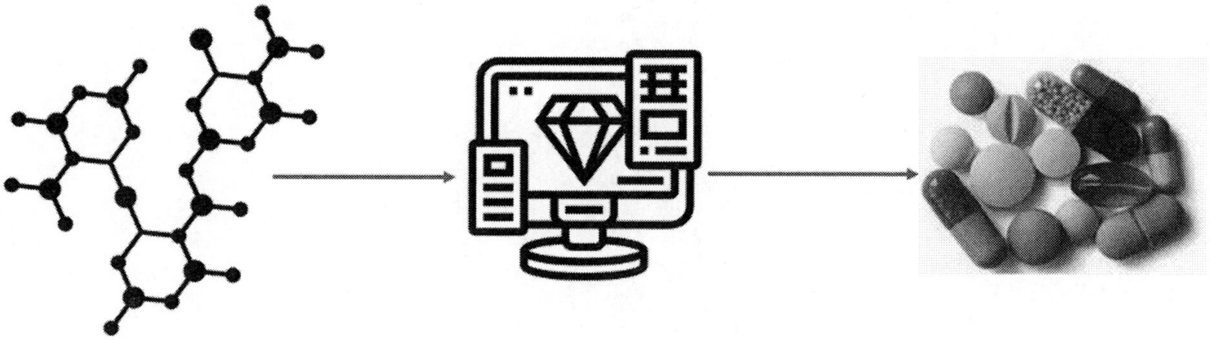

606b Identifying and Addressing the Weaknesses of Density Functional Theory for Crystal Structure Prediction

Gregory Beran

Density functional theory has enabled major advances in pharmaceutical crystal structure prediction. Despite these successes, however, there are challenges. Delocalization error inherent in the approximate functionals can bias crystal structure predictions toward certain types of structures. We have found quite a few examples of this problematic behavior in describing the intramolecular conformational energies, for example. Descriptions of the intermolecular interactions in the lattice can also be problematic. Examples of these issues and strategies for addressing them will be discussed.

606c Rational Reduction of Computationally Predicted Crystal Energy Landscapes Using Molecular Dynamics and Enhanced Sampling Techniques.

Nicholas F. Francia, Louise Price, Sally L. Price, Matteo Salvalaglio

Crystal Structure Prediction (CSP) methods should ideally identify the operative conditions to obtain a specific polymorph starting only from the molecular structure of their building blocks. Unfortunately, even if CSP methods have dramatically evolved over the past decade, the field is still far from achieving this vision.

While successful predictions of the most stable polymorphs of organic crystals based on accurate lattice (free) energy estimates are becoming increasingly common, the number of predicted putative polymorphs remains a standing issue.
Typically, the overall number of potential structures computationally identified as putative polymorphs usually grossly exceeds the number of polymorphs observed experimentally [1]. This problem in the field is typically referred to as 'overprediction'.

A significant reason for CSP methods over-predicting possible polymorphs is that temperature increases molecular motions so that some structures that appear as distinct lattice energy minima merge into the same free energy minimum at finite temperature [2].
This work discusses applying a physics-driven computational procedure [3,4] to reduce the number of relevant local minima of crystal energy landscapes based on a systematic and scalable application of molecular dynamics and enhanced sampling simulations.

In order to identify persistent crystal structures, we perform classical molecular dynamics simulations at finite temperature on CSP-generated crystal structures. Unstable configurations are then automatically removed by comparing the distribution of internal orientations against the random arrangement typical of a disordered melted state.
To identify crystal structures that convert to a common finite-temperature state, we carry out a clustering analysis based on probabilistic fingerprints that capture information on the relative position, orientation, and conformation of molecules within a dynamic crystal supercell. Differences in the probabilistic fingerprints concur to define a dissimilarity metric between finite-temperature structures, which is then used to perform unsupervised clustering of finite temperature structures.

In this contribution, we discuss the application of this method to systems of increasing size and complexity (urea, succinic acid, ibuprofen, olanzapine and others), spanning from a few dozens to thousands of structures. We note that, in all cases, we substantially reduce the crystal energy landscapes while consistently retaining the experimentally observed crystal structures.
Developing a Python library [5] that manages the setup of MD simulations and automatically analyses the resulting trajectories has been instrumental in achieving scalability over a large set of crystal structures, enabling us to apply MD workflows to sets of crystal structures approaching the size and complexity of real-world CSP applications.

References

[1] DA Bardwell, CS Adjiman, YA Arnautova, E Bartashevich, SXM Boerrigter, et al., 2011, Acta Cryst. B 67:535-55
[2] Price SL. 2013, Acta Cryst. B. 69:313-28.
[3] Francia, N.F., Price, L.S., Nyman, J., Price, S.L. and Salvalaglio, M., 2020. *Crystal Growth andamp; Design*, *20*(10), pp.6847-6862.
[4] Francia, N.F., Price, L.S. and Salvalaglio, M., 2021, *CrystEngComm*, *23*(33), pp.5575-5584.
[5] https://github.com/mme-ucl/pypol

Figure: Reduction of the crystal energy landscape of Urea. In the outer circle the lattice energy minima, while in the inner circle the 12 structures that survived the reduction process. In light grey the structures that melt while in different colours those that coalesce to the same crystal structure.

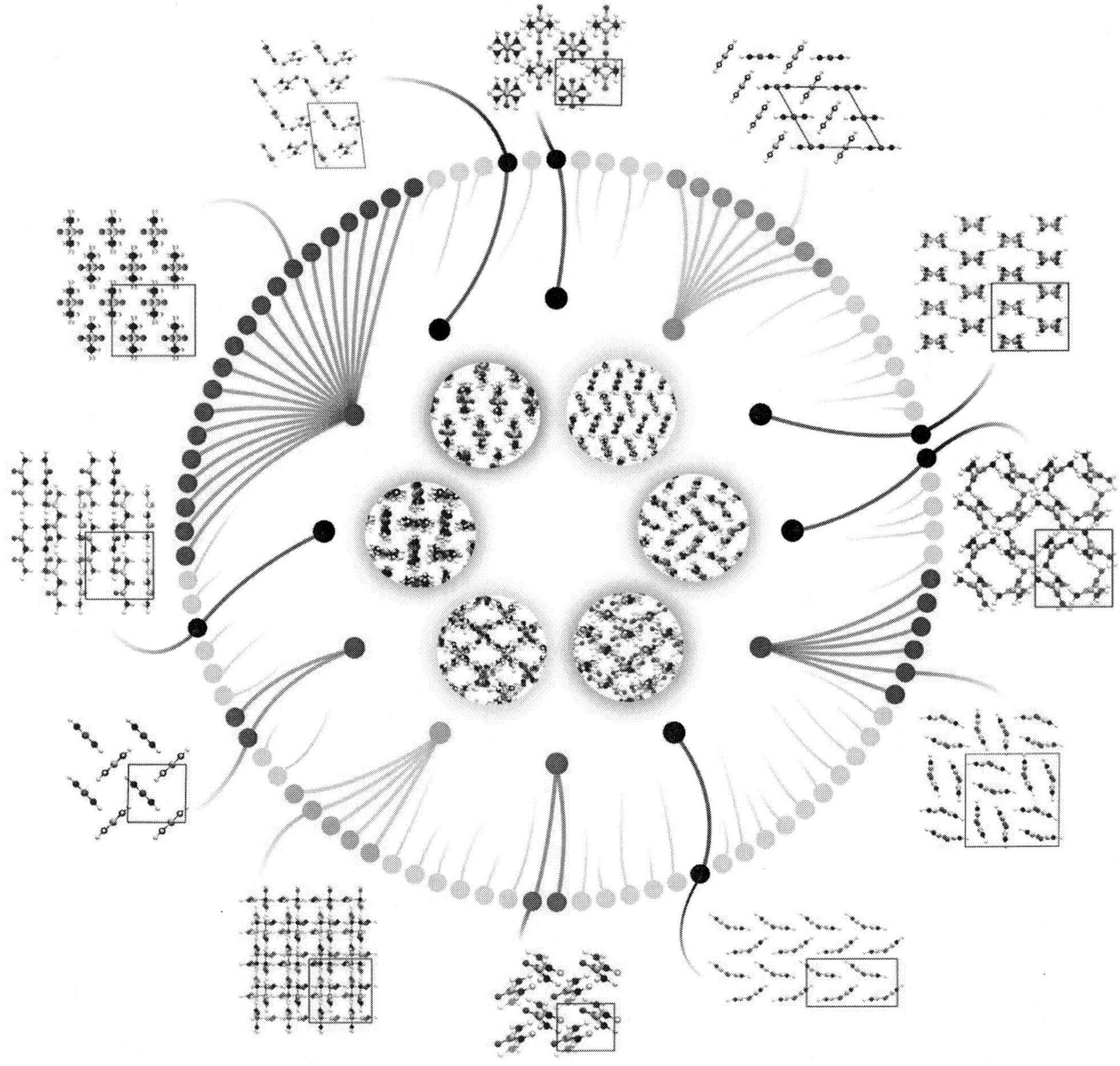

606e A Novel Computational Approach to Guide Impurities Rejection By Crystallization

Yuriy Abramov, Amanuel Zelellow, Michal Achmatowicz, Cheng Chen, Jian Wang, Sivakumar Sekharan, Michael A. Bellucci

Process organic impurities arise from many sources throughout the synthesis and storage of active pharmaceutical ingredients (APIs). These impurities may include the starting materials, intermediates, byproducts, reagents, and degradation products. Regulatory expectations for control of impurities in new drugs have been established through ICH guidelines, which outline requirements for the registration of new drugs. The most efficient approach to impurity rejection and control is provided by API (re)crystallization. Until recently, molecular modeling support was focused predominantly on rational solvent selection for impurity rejection, considering one impurity at a time in one solid form of API.[1] Here we present a new model that allows assessment and comparison of multiple impurities in parallel, and describes the effects of initial impurity concentration (or doping), as well as API solid form and impurity-lattice interaction on impurity rejection. The model was successfully tested on multiple public domain examples. In addition, a special validation study was designed for impurities rejection of an investigational drug candidate.

1. Abramov, Y.A. Rational Solvent Selection for Pharmaceutical Impurity Purge. *Cryst. Growth Des.* **2018**, 18, 1208−1214.

606f Polymorphism in a Lennard-Jones Fluid Nucleating from Its Melt

Pelin Su Bulutoglu, Shiyan Wang, Nandkishor Nere, Doraiswami Ramkrishna

Polymorph selection, which is predicting the outcome of a crystallization process in terms of the resulting crystal structure, has important applications in fine chemicals industries. For example, polymorphs of active pharmaceutical ingredients have different solubilities and dissolution rates resulting in varied bioavailability of drugs. Computational prediction of the kinetically favored polymorph can reduce experimental costs significantly and accelerate formulation development. In this work, we take an atomic-resolution look into the competition between HCP and FCC polymorphs in a Lennard-Jones (LJ) fluid nucleating from its melt using molecular dynamics (MD) simulations and 2-dimensional free energy calculation with hybrid Monte Carlo/Molecular Dynamics (HMC/MD) simulations. Structural composition of the nucleus is considered as well as nucleus size by calculating the free energy as a function of structure specific nucleus size coordinates. Unbiased MD trajectories reveal that polymorph selection happens after nucleation of an FCC-dominated cluster, also confirmed by the presence of a single free energy saddle point on the polymorph specific nucleus size space. The critical cluster transforms into either FCC or HCP phases in the growth stage. By calculating the structure specific diffusion coefficients through unbiased MD simulations, we construct the complete mesoscopic system for LJ nucleation from the melt and calculate nucleation rates. The choice of LJ fluid as a model system is advantageous in elucidating the strategy for formulating a mesoscopic model for calculating nucleation rates of multiple structures, which can be applied to the more computationally intensive problem of real systems with more complex molecules. Direct comparison of computational predictions with experimental measurements is challenging for polymorph specific nucleation rates since monitoring the structure of the critical nuclei is difficult and solvent mediated polymorphic transitions can cause errors in the measured rates. We propose a population balance model that captures the nucleation, growth, and polymorphic transformations of a population of clusters in LJ nucleation from the melt. This multi-scale, first-principles based model can potentially predict the polymorphic outcome of a crystallization process and allow a direct comparison to experiments.

606g Assessment of Force Fields for Crystal Morphology Prediction

Yongsheng Zhao, Robert Gee, Michael F. Doherty

Crystal morphology plays a significant role in many industrial applications, and it is a challenging task for the industry to obtain desired crystals with particular product functionality and morphology. Therefore, our group at UCSB has developed an academic design software aid called ADDICT (Advanced Design and Development of Industrial Crystallization Technology) to predict crystal morphology based on mechanistic models [1]. Recently, we have upgraded ADDICT's software framework [2] and completely rewritten the code based on Matlab using object-oriented programming in ADDICT3 (the third version of ADDICT) [3].

The interaction energy between different growth units is the key to the mechanistic growth models. Its accuracy directly determines the quality of the prediction results. Different force fields may result in different interaction energies, which further lead to different crystal morphologies. Therefore, it is very helpful to study the effect of different force fields on the morphology of crystals. Generalized Amber Force Field [4] (GAFF 1.8) is the first force field to be embedded in ADDICT. Recently, we also added the Universal force field (UFF) [5], Coulomb–London–Pauli (CLP) [6], and Lifson [7] force fields. Based on the above force fields, ADDICT can be used for predicting the morphology of various organic crystals containing a wide variety of atom types.

In this work, we systematically studied and compared the effect of different force fields in ADDICT on the morphology of several crystals (e.g., naphthalene, adipic acid, biphenyl, pentaerythritol, olanzapine, β-polymorphs of copper phthalocyanine, etc.). We also discuss the effect of the Hybrid Force Field (HFF) based on the combination of these different force and solvent effect (e.g., using our own versions of COSMO-SAC) on the crystal morphology. The results demonstrate that different force fields have different advantages and scope of application. For example, the Lifson force field can be used to predict very well the morphology of carboxylic acid crystals, etc. In addition, we also discuss the effect of the optimization of the H atom position on the crystal morphology.

References

[1] Li, J., Tilbury, C. J., Kim, S. H., andamp; Doherty, M. F. (2016). Prog. Mater. Sci., 82, 1.

[2] Zhao, Y., Tilbury, C. J., Landis, S., Sun, Y., Li, J., Zhu, P., andamp; Doherty, M. F. (2020). Cryst. Growth Des., 20, 2885.

[3] Landis, S., Zhao, Y., andamp; Doherty, M. F. (2020). Comput. Chem. Eng., 133, 106637.

[4] Wang, J., Wolf, R. M., Caldwell, J. W., Kollman, P. A., andamp; Case, D. A. (2004) J. Comp. Chem., 25, 1157.

[5] Rappé, A. K., Casewit, C. J., Colwell, K., Goddard III, W. A., andamp; Skiff, W. M. (1992). J. Am. Chem. Soc., 114, 10024.

[6] Gavezzotti, A. (2011). New J. Chem. 35, 1360.

[7] Lifson, S., Hagler, A. T., andamp; Dauber, P. (1979). J. Am. Chem. Soc., 101, 5111.

272b Digital Twin Development for Continuous Drug Manufacturing

Busuyi Adebayo, Syed Ahmed, Rob Guenard

Rising above numerous setbacks and inherent challenges, biopharmaceutical industry has started to consider the integration of upstream and downstream operations for continuous drug manufacturing to shore up drug supply chain, reduce (cross)contamination, and time to market to mention but a few. For successsful development of this in intricate processes of drug manufacturing, mechanistic modeling and simulation has been recommended by subject matter experts, policy makers and regulation agencies such as FDA and EMA.

Herein, we focus on:

1. Mechanistic modeling and simulation of the unit operations in a typical continuous drug manufactuirng train

2. Model validation with experimental data

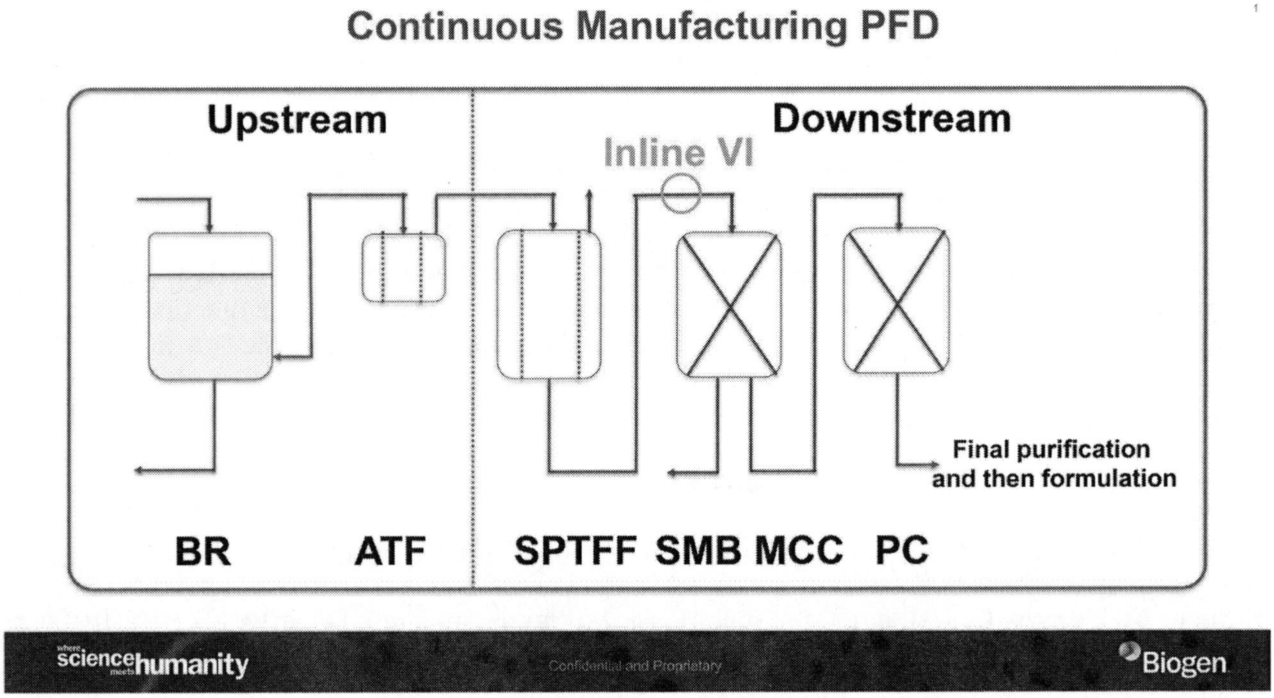

272d Integrated Flowsheet Model of Continuous API Manufacturing Process and Its Applications for Dynamic Optimization and Control

Ravendra Singh, Fernando Muzzio

Advanced continuous manufacturing (CM) is emerging as a preferred platform to produce active pharmaceutical ingredients (APIs). The future continuous pharmaceutical manufacturing is desired to be based on modular plug and play concept. In this approach, mobile modules consisting of a group of unit operations have been developed that can be easily reconfigured to assemble a scalable pharmaceutical manufacturing process for different products. It is also suitable for on-demand manufacturing such as in the case of COVID-19. However, because of different levels of complexity, the development and adaptation of flexible and modular CM plant is still an open area of research. Therefore, a systematic framework of methods and tools are needed to develop, adapt and evaluate the flexible modular CM plant for API manufacturing with reduced time and resources. In particular, a digital twin is critical for quick design, adaptation, optimization, scale up, and control of continuous APIs manufacturing process, allowing for virtual modeling of operations without extensive process development.

In this work, a digital twin for continuous APIs manufacturing process using modular components has been developed and its applications have been demonstrated for dynamic optimization and control system design. The process flowsheet model is the heart of 'digital twin' and consists of mathematical representation of three modules with the options of adding more as needed. The first module is for feeding and dispensing. It consists of a refill unit, feed tank, pump, mixing, and preheater. The second module is for performing the chemical reactions required to produce the target APIs. This module consists of tubular reactors placed inside a heating and cooling jacket. The third module is focused on separation of API from impurities. A continuous chromatographic model is currently used for separation purposes that can be easily replaced with any other type of separation techniques such as continuous crystallization. Currently, the developed model library consists of 32 units across 10 types that has been used to generate the integrated flowsheet model. The model has been validated using experimental data.

The applications of the digital twin have been demonstrated for dynamic optimization as well as design and development of suitable control architecture for a continuous APIs manufacturing process. The dynamic optimization provides the optimum operating conditions for the manufacturing process that improves the critical quality attributes (CQAs). A suitable control architecture then has been developed using digital twin to track the optimum set points for the key control variables. The developed control architecture is also modular in nature and can be easily adapted for different manufacturing processes of APIs that may have different material and information flows, dead time variations, and tuning parameters. The proposed digital twin can save the time and resources needed for continuous APIs manufacturing and improve the product quality significantly.

The objective of this presentation is two-fold; first to highlight the developed unit operation model library and integrated flowsheet model and then demonstrate its application for dynamic optimization and design of control architecture of continuous API manufacturing processes.

272e Process Development and Scale-up of Continuous Hydrogenation from Lab Scale to Pilot Plant in a Trickle Bed Reactor

Onkar Manjrekar, Christopher Vitale, Greg Storer, Brian Kotecki, Narendrabh Patel, Duygu Gerceker, Jie Chen, Moiz Diwan

In past decade there has been interest in pharmaceutical industry on development of continuous hydrogenation processes. In this work lab scale development of continuous hydrogenation process in trickle bed reactor and scale up to pilot scale is described. Fit for purpose kinetic model is developed for scale-up in GMP-setting, implementation of online HPLC for reaction monitoring, and consideration of equipment to operate in class-1 division-2 area would be discussed.

272f Tracking Lot Genealogy in Continuous Drug Substance Manufacturing Processes through Mechanistic Modeling

Meera Mahadevan, Dana Barrasso

Due to the nature of continuous processing, varying lot numbers of each given material can be present in a continuous drug substance manufacturing (CDSM) operation at any given time in various concentrations. Tracking all the lot numbers throughout the process is a challenge due to variations in flow rates, lot changes, dispersion in plug flow reactors, one or more steps not being purely continuous, and process disturbances which result in variations in residence time distributions. Despite these challenges, key drivers for tracking lot "concentrations" throughout a CDSM process include:

- Regulatory quality – to prove genealogy in the system for batch release
- Batch definition in production campaigns (reduce bottlenecks in release process for longer runs)

In this talk, an approach will be presented using mechanistic modeling to track lot composition throughout CDSM processes from the feed tanks to isolation using residence time distribution (RTD) models and flowsheet modeling methodologies. This framework will be demonstrated on a realistic CDSM process, quantifying the presence of each lot in the final product. Scenarios related to feed tank operational decisions and subsequent processing steps will be explored.

Such a framework may also be used to monitor lot composition as part of a process's automation system, aligning with the industry's efforts to ensure regulatory quality in continuous manufacturing operations. This topic will also be discussed in the presentation.

272g A Step-By-Step Strategy for Modeling, Scaling-up, and Control of Aspirin Production in a Continuous Tubular Antisolvent Crystallizer Under Ultrasound

Symeon Savvopoulos, Mohammed N. Hussain, Spyridon Voutetakis, Tom Van Gerven, Dimitris Ipsakis, Simon Kuhn

The crystallization of organic nano and microparticles is an important process for several industries, products, and applications [1]. The formed particles are crucial for various applications, such as in the case of organic micron-sized crystals of active pharmaceutical ingredients (APIs) and nanodispersions of coating resins and microcapsules with active ingredients [1]. The crystallization process itself can be performed in batch, semi-batch or continuous mode. Currently, batch crystallization is widely implemented at a large scale to meet market needs. However, batch processes are facing persistent challenges during operation. An indicative example is a hurdle posed during the control of heat and mass transfer; two factors that affect both the average particle size and particle size distribution (PSD), as well as the batch-to-batch variability. Continuous crystallization comes as a promising initiative to enable the maintenance of higher product standards [2].

Continuous processes allow the production of particles with specific particle size, polymorphism and smaller PSD [3]. Additionally, continuous processes produce quantities on demand, minimizing the use of raw materials and equipment. Furthermore, mixing energy and quantity of solvent could be decreased, and, last but not least, they can be scaled up easier as compared to batch systems [4]. Ultrasound can be used for improving continuous crystallization processes because the generated bubbles act as cluster attachment sites, forming finer crystals with a small average size and narrower PSD. However, the nucleation mechanism that takes place during the ultrasound is still unclear [5].

In this work, initially we demonstrate the model development of the continuous antisolvent crystallization of aspirin in water and ethanol in an experimental plug flow crystallization with ultrasound, and its validation with experimental measurements [6]. Concerning the mathematical modeling, population balance models (PBMs) are used extensively because they are available to capture the changes in the number and size of crystals over time, and space. Heat generation due to mixing and ultrasound, which increases crystal solubility and affects the yield of the system subsequently, is also included in the model. A previously validated model is then used for both theoretical scale-up and control. Regarding the scale-up, a CFD model is developed and validated in an effort to study the ultrasound-assisted continuous antisolvent crystallization of aspirin in different dimensions. Therefore, the first tubular crystallizer model is coupled with a validated acoustic field in the experimental set-up which is generated by the ultrasound transducer. The acoustic field in the crystallizer is identified experimentally with sonochemilluminescence experiments. The next part is the consideration of the proper fluid flow model. Although the experimental flow is laminar, the ultrasound bubbles cause turbulent and improved micromixing conditions. The fluid flow pattern is selected by comparing the CFD simulations in both laminar and turbulent flows with the results of the 1-D model by keeping the same previous expressions of solubility, nucleation, and growth rates. The accurate CFD model is used to predict the crystal growth in the presence of ultrasound at an increased diameter [7].

Finally, the last part is focused on a control concept and strategy for the optimization of the volume-based mean crystal size of the PSD. Toward this direction, this work's core objective is to develop and assess a conventional feedback controller for the ultrasound-assisted continuous antisolvent crystallization of aspirin. In this way a simple, and accurate, controller scheme that controls the volume-based mean crystal size under disturbances in both the active pharmaceutical ingredient inlet concentration, and the inlet temperature (taking into consideration different profiles of disturbances) is presented. During simulated case scenarios, the control scheme is evaluated in extreme conditions of disturbances in the solvent flow rate [8].

This step-by-step strategy, which combines experiments and mathematical modeling, is crucial in a) capturing the crystal growth in continuous crystallizers under the effect of ultrasound, b) studying the continuous crystallizers scale-up with ultrasound, and c) applying control. Subsequently, the steps presented will provide insight into the design of new crystallizers that operate in continuous mode.

References

1. Gao Z, Rohani S, Gong J, Wang J. Recent Developments in the Crystallization Process: Toward the Pharmaceutical Industry. Engineering. 2017;3(3):343–53.
2. Yang X, Acevedo D, Mohammad A, Pavurala N, Wu H, Brayton AL, et al. Risk Considerations on Developing a Continuous Crystallization System for Carbamazepine. Org Process Res Dev. 2017;21(7):1021–33.
3. Sang-Il Kwon J, Nayhouse M, Orkoulas G, Christofides PD. Crystal shape and size control using a plug flow crystallization configuration. Chem Eng Sci. 2014;119:30–9.
4. Da Rosa CA, Braatz RD. Multiscale Modeling and Simulation of Macromixing, Micromixing, and Crystal Size Distribution in Radial Mixers/Crystallizers. Ind Eng Chem Res. 2018;57(15):5433–41.
5. Jordens J, Canini E, Gielen B, Van Gerven T, Braeken L. Ultrasound assisted particle size control by continuous seed generation and batch growth. Crystals. 2017;7(7).
6. Savvopoulos S V., Hussain MN, Jordens J, Waldherr S, Van Gerven T, Kuhn S. A Mathematical Model of the Ultrasound-Assisted Continuous Tubular Crystallization of Aspirin. Cryst Growth Des. 2019;19(9):5111–22.
7. Savvopoulos S V., Hussain MN, Van Gerven T, Kuhn S. Theoretical Study of the Scalability of a Sonicated Continuous Crystallizer for the Production of Aspirin. Ind Eng Chem Res. 2020;59(45):19952–63.
8. Savvopoulos S V., Voutetakis SS, Kuhn S, Ipsakis D. Theoretical Feedback Control Scheme for the Ultrasound-Assisted Continuous Antisolvent Crystallization of Aspirin in a Tubular Crystallizer. Ind Eng Chem Res. 2021;60(17):6221–34.

138a Development of an Adsorbent Treatment Workflow for Impurity Removal

Zhenshu Wang, Saurin Hiren Rawal, PhD, Jeremy Merritt, Daniel J. Jarmer

Critical Quality Attributes (CQAs) of a small molecule drug substance, such as chemical purity and physical properties, are often correlated due to the impact of impurities on both crystal morphology and crystallization kinetics.[1] Addressing the CQAs with a robust crystallization process can be difficult when impurities cause liquid-liquid phase separation (oiling) during crystallization.[2] Adsorbent treatments such as functionalized silica or activated carbon can be used to purify process streams and reduce the occurrence of oiling during crystallization.[3-5] For this case study involving a small molecule drug substance, an adsorbent treatment workflow was developed to determine when an adsorbent treatment is needed and to facilitate process design. Key design parameters investigated include temperature, solvent system, adsorbent type, and adsorbent loading. A partial least squares (PLS) regression model was developed to evaluate the data, and a database was created for future projects to enable rapid statistical analysis and to correlate absorption performance (*e.g.,* selectivity, yield) to drug substance chemical purity and physical properties.

1. Urwin, S. J., Yerdelen, S., Houson, I., andamp; ter Horst, J. H. (2021). Impact of impurities on crystallization and product quality: a case study with paracetamol. *Crystals*, *11*(11), 1344.
2. Meng, Z., Huang, Y., Cheng, S., andamp; Wang, J. (2020). Investigation of Oiling-Out Phenomenon of Small Organic Molecules in Crystallization Processes: A Review. *ChemistrySelect*, *5*(26), 7855-7866.
3. Daver, S., Rodeville, N., Pineau, F., Arlabosse, J. M., Moureou, C., Muller, F., ... andamp; Cardinaud, I. (2017). Process development and crystallization in oiling-out system of a novel topical antiandrogen. *Organic Process Research andamp; Development*, *21*(2), 231-240.
4. Adlington, N. K., Black, S. N., andamp; Adshead, D. L. (2013). How to use the lasentec FBRM probe on manufacturing scale. *Organic Process Research andamp; Development*, *17*(3), 557-567.
5. Welch, C. J., Albaneze-Walker, J., Leonard, W. R., Biba, M., DaSilva, J., Henderson, D., ... andamp; Wang, T. (2005). Adsorbent screening for metal impurity removal in pharmaceutical process research. *Organic process research andamp; development*, *9*(2), 198-205.

138b Development and Control of a Late-Stage API Crystallization Via Kinetics Modeling and Controlled Crystallization

Shujauddin Changi, Berenice Lewandowski, Satyarit Rao, Billie Kline, David Willcox

To supply the early developmental batches for clinical trials, a fit-for-purpose API process step included an ammonolysis reaction followed by uncontrolled crystallization – involving crash-out of the API due to its poor solubility in the system. This crude crystallization was followed up by a recrystallization step in a different solvent matrix – allowing for polish filtration and reprocessing options. The impurity purging with either the crude or the recrystallization steps were not great. As the program advanced, an alternate process was developed with an aim to streamline, robustly scale-up, and help establish impurity control. The new process utilized a kinetics model to enrich the reaction understanding and ensured that a controlled crystallization was possible post reaction that allowed for a better impurity purging. This talk aims to focus on the developmental aspects of this journey, highlighting the motivation, challenges, and process improvements for the same.

138c Control of Ritter Side Products Via Reaction Engineering and Kinetic Modeling to Enable Robust Scale-up

Jennifer Lott, Antonio C. Ferretti, Geoffrey E. Purdum, Antonio Ramirez, Scott Bader, Hon-Wah Man, Gerald Artman

The final transformation step for an API requires the installation of a glutarimide ring using a t-butyl deprotection/cyclization sequence. The release of isobutylene from the t-butyl group during the reaction leads to the formation of impurities via the Ritter reaction. Reaction engineering efforts were undertaken to minimize these impurities through mechanistic understanding and process optimization. Kinetic profiling experiments provided insight into the fate of the t-butyl cation through formation of the impurities, consumption by the solvent acetonitrile, and evolution of isobutylene in the gas phase. A kinetic model was developed, enabling prediction of the impurities and side product with acetonitrile. The reaction was optimized leveraging temperature and dilution as scale-independent parameters in combination with nitrogen sparging. The use of fundamental data and modeling enabled the development of the control strategy for this final transformation of the API.

Title: **Wet milling of mefenamic acid for seed generation: model-driven size reduction for maximizing yield**

Mehta, B. A.[1,2], Brown, C. J.[2], Mitchell, N. A.[1], Florence, A.[2]
[1]*Siemens Process Systems Engineering Ltd., London, UK*
[2]*EPSRC Future CMAC Manufacturing Hub University of Strathclyde, Glasgow, UK*

Keywords: wet milling, breakage, growth, design space exploration

Introduction:

There is an increasing interest in the application of continuous processing technologies using continuous crystallization in pharmaceutical manufacturing to control crystal properties and deliver consistent particulate products (Bosetti & Mazzotti, 2020; Brown et al., 2018). Whilst widely used in the production of large volume commodity chemicals, there is increasing number of industrial examples and support from regulatory authorities including the FDA in demonstrating the potential value of continuous crystallization in the supply of medicines and other high value chemicals. Recent integration of wet media milling technologies, particularly rotor-stator devices, have been shown to avoid common issues encountered with traditional dry milling units. These include introduction of crystal lattice disorder, undesired polymorphic and amorphous transformations, reduced operator exposure to high potency APIs, broad bimodal particle size distributions, loss of yield and process control, as well as increased energy cost (Agimelen et al., 2018; Ahmed et al., 2019; Szilagyi & Nagy, 2019).

The aim of this work is to explore the region of attainable particle sizes based on the performance of a wet milling model. In detail, this work focuses on the development of a mechanistic model to describe the wet milling process, with a view to parametrize the breakage kinetics, optimize the milling performance and explore the effects on downstream processes, such as crystallization.

Materials and Method:

Equipment and Material specifications

Mettler Toledo OptiMax™ workstation consisting of a 1 L stirred tank crystallizer (STC) equipped with an Hastelloy® Pt100 temperature sensor was used for all experiments with the system controlled and operated through the iControl v5.4 software. The following Mettler Toledo *in situ* process analytical technology tools used for each of the experiments were: FBRM probe (G400 series) for particle tracking and chord length distribution by laser back-scattering (iC FBRM V4.3) and a PVM V819 probe with online image acquisition software V8.3 included. An IKA MagicLab (Module UTL) wet mill unit was used for the inline milling process. Mill outlet temperature was monitored on the IKA operating unit display. Rotational speeds were operated from 3000 to 26000 rpm, with varying teeth configurations investigated across coarse, medium and fine rotor-stator tooth pairs. The temperature of the wet mill chamber was manually controlled during milling to match the temperature of the feed solution within 1 °C throughout cooling. The API and solvent system were the same as used for crystallization, that is Mefenamic Acid in 90:10 w/w 2-butanol and n-heptane.

Initial Model Configuration

A single wet mill experiment was conducted to identify the smallest possible particle size that could be achieved from the above configuration. The slurry was prepared with 90g of API and 702g of the saturated solution, making up 1L. rotor-stator mill with a fine generator was run at a rotational speed of 26,000 RPM for 1 hour with a mill recycle flow rate of 705.6 g/min.

The initial model was developed using gPROMS FormulatedProducts 2022.1 using an off-the-shelf wet milling population balance model. The breakage kinetics is given below (Austin, Shoji, Bhatia, Jindal, & Savage, 1976):

$$B_M(v,\omega) = \phi\left(\frac{v}{\omega}\right)^{\frac{\gamma}{3}} + (1-\phi)\left(\frac{v}{\omega}\right)^{\frac{\beta'}{3}}$$

$$S_M(\omega) = \begin{cases} Sc_M\left(\dfrac{\omega}{\omega_{critical}}\right)^{\frac{\alpha'}{3}} & for\ \omega \geq \omega_{critical} \\ 0 & for\ \omega < \omega_{critical} \end{cases}$$

γ, β', ϕ and α'	Kernel parameters	-
w and y	Breaking particle volumes and corresponding sizes	m^3 and μm
v and x	Product particle volumes and corresponding sizes	m^3 and μm
$w_{critical}$ and $y_{critical}$	Critical volumes and corresponding sizes that prevent particle from further breakage	m^3 and μm
Sc_M	Constant rate of breakage at critical size	s^{-1}

The linear correlation between breakage rate and shear rate is given by the following expression (Karabelas, 1976):

$$S = C\gamma$$

For rotor-stator wet mills, the equation can be generalised to include gap thickness between the rotor and stator, and the rotor frequencies:

$$S = C\sum_{i=1}^{N_{\gamma-\delta}} \frac{v_{ts,t}}{\delta_i} + C_0$$

v_{ts}	Rotor/impeller tip speed	m/s
δ	Horizontal gap between the rotor and stator	m
C_0	Intercept parameter, accounts for additional breakage that may occur as the tip speed tends to zero	1/s
N_{r-s}	Number of rotor-stator rows	-

In-situ chord length measurements were taken from the crystallizer vessel. These measurements were converted to volume mean lengths and were used to estimate the kinetic parameters in the above equations. The critical particle size (C_0) was fixed at the smallest value recorded at the end of the experiment.

Additional experiments

Based on the fit from the single experiment, further lab work was suggested with a similar method as that described above. The key differences in the new experiments were:

(a) A coarse rotor-stator mill setup was suggested. These experiments would allow for the re-validation of the equipment correlation, and thus allow for a more robust design space exploration.
(b) Each experiment was done at different rotational speeds, varying between 3000 to 26000 RPM, which would lead to a wider design space study

Design Space Exploration using Global Systems Analysis

The main use of process simulation and modelling tools is to analyse "what-if" scenarios in order to improve process design and operation. In the absence of Global System Analysis, this is done manually as a point activity, using repeated simulation runs. This is very onerous and gives only a very limited view of the actual decision space.

Global System Analysis ('GSA') allows users to comprehensively explore the behaviour of a system over domains of any user-selected subset of its input variables ('factors'), and output variables ('responses'), and analyse results in a comprehensible way. This provides a quick, easy and systematic way to explore the complex process design and operational decision space using high-fidelity models.

The primary objective of this GSA study was to obtain a region of attainable particle sizes. This would lead to a better understanding of mill characteristics, and hence reduce the need for trial and error in the lab to get the required particle size.

<u>Results</u>

Initial Model Development

Figure 1 and Table 1 below show the main results from the parameter estimation. It can be seen that the fit for the prediction vs the experimental data is generally very good and fits within the experimental errors. From Table Y, values for each estimated parameter and the objective function are given. The objective function is a value that directly quantifies how good the predictions are when compared to the experimental data. Generally, the lower the objective function, the better the fit.

Table 1: Parameter estimation results

Equipment Correlation Parameter – C_0	Equipment Correlation Parameter – C	Critical Particle Size at which particles no longer break	Kapur Parameter (q)	Austin Parameter (α)	Objective Function
0.349332	0	1.97587×10^{-8}	45	0.883269	7.33086

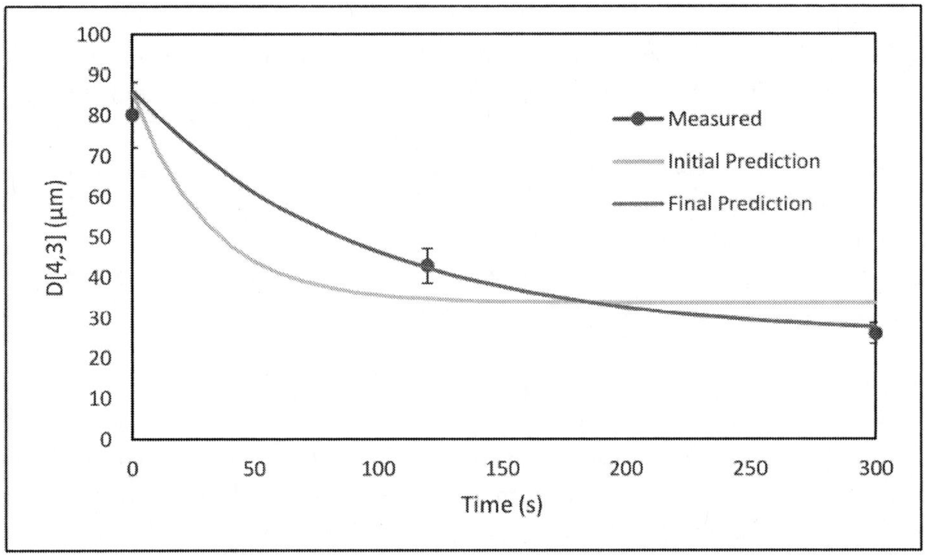

Figure 1: Measured vs predicted volume mean particle sizes for the wet mill experiment

Further Experiments, Revalidation, and Design Space Exploration

The design of further experiments was done using a rough GSA analysis. The results from the GSA are shown in Figure 2. Do note that the model currently is not able to distinguish between different rotor-stator generators, so the results show the variation of particle size vs time with varying rotational speeds for the fine generator.

GSA results suggest that lower milling rates would potentially lead to a more gradual decrease in particle size. This can be predicted without GSA, however the use of GSA shows that a majority of the milling occurs in the first 15 minutes of the experiment and that regardless of rotational speed, there is very little variation in particle size after milling for 30 minutes (indicated by the black dashed line in Figure 2. This is useful information because a data collection procedure could then be defined based on this information. Table 2 summarizes the suggested experiments that were carried out in the lab, along with the data collection procedure and key results.

Table 2: Summary of further experiments performed

API	Mefenamic Acid
Solution	90:10 w/w 2-butanol:heptane
Solution mass (saturated) (g)	702g
API load (g)	90g
Mill generator type	Coarse
Total number of teeth (rotor + stator)	13
Rotational speeds (RPM)	3000-26000
Number of Experiments	6

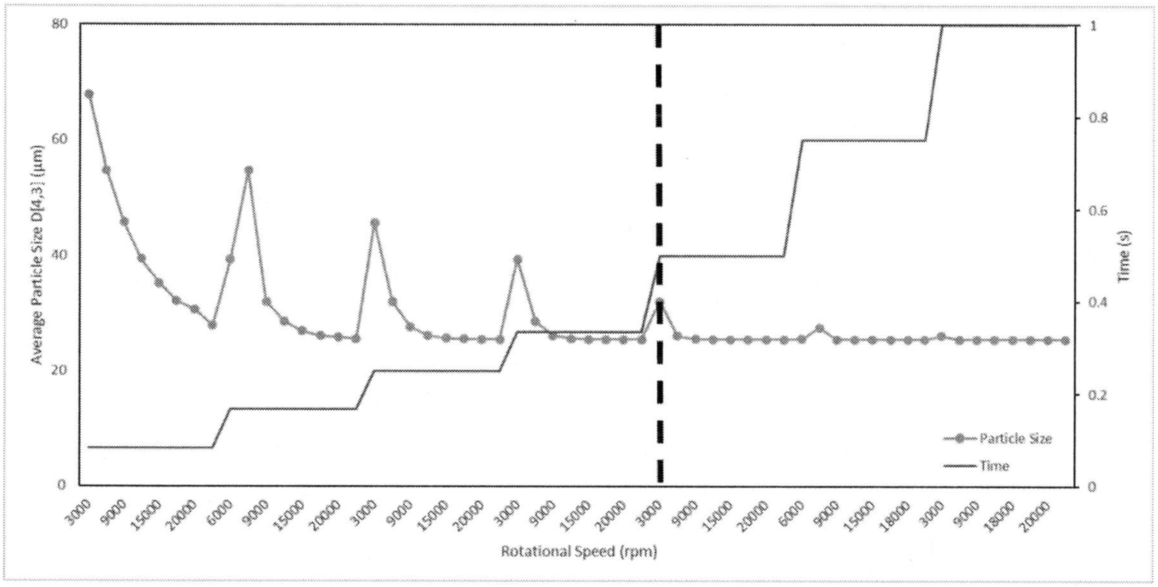

Figure 2: Design space exploration results for particle size variation over time and with varying rotational mill speeds

The new experiments were added to the model, and a validation activity was carried out to compare how good the existing parameters were at predicting the experimental outcome. The results, displayed in Figure 3, show that the fit for the higher rotational speed experiments were good, even though a different mill generator was employed in the new experiments (coarse generator for new vs fine generator for the old experiment). From these results, a more rigorous GSA study was done to obtain a region of attainable sizes and thus suggest what lab-scale experiment is the most appropriate and most efficient in achieving the desired d90.

Figure 4 displays the region of attainable sizes split by generator. The dark blue area of the plots below shows the desired particle size, that is a d90 of 42µm. One key observation from the graph below is that any of the generators can provide the desired particle sizes, with the fine generator achieving this in a shorter time frame as compared to the medium and the coarse generators. Furthermore, the fine generator had the highest probability of achieving the required d90 with a shorter experimental run.

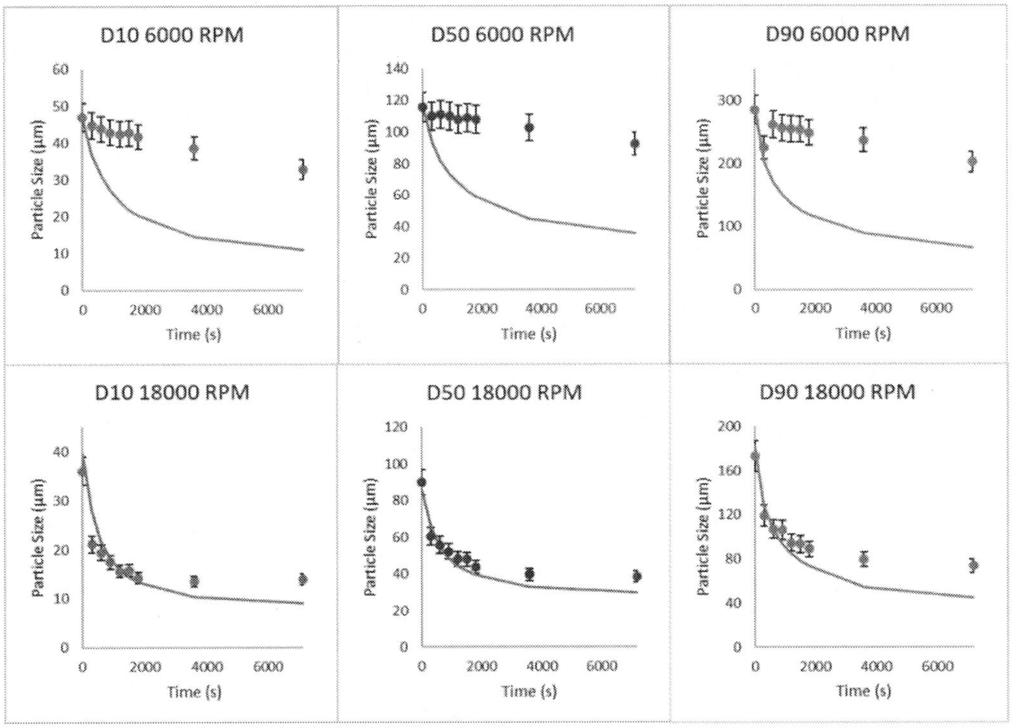

Figure 3: Results from model validation using further experiments for D10, D50 and D90 PSD quantiles

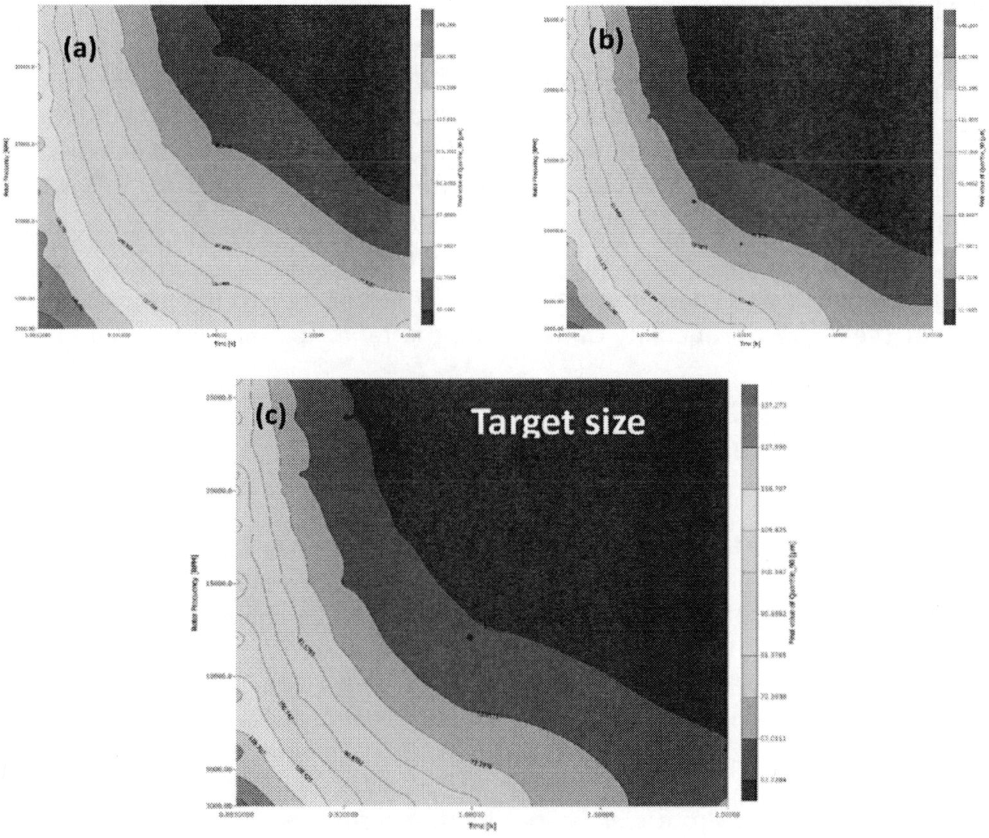

Figure 4: Region of attainable particle sizes for (a) coarse, (b) medium, and (c) fine rotor stator configurations

Ultimately, the final lab-scale experimental suggestion for achieving the required d90 was chosen based on the following criteria:

(a) Highest probability of achieving the required particle size
(b) The length of experiment should factor in any potential experimental error
(c) Most efficient rotor-stator configuration based on energy dissipation and number of rotor-stator teeth

From these requirements, the suggested run was set at using the fine rotor-stator mill configuration with rotational speeds between 18000 – 26000 RPM for a runtime of 60 minutes.

Conclusion

This work demonstrated a model-based design approach for design and development of a wet milling process for seed generation. A single experiment was used to estimate breakage kinetics in a wet mill mechanistic model. The validated model was used to explore the design space and suggest further experiments with different rotor-stator configurations. The new experiments were shown to fit well with earlier estimated breakage kinetics, despite the change in mill configuration. Predictions were good for both the fine and coarse generators, prompting a wider design space exploration for different mill configurations. GSA results showed that the highest probability of achieving the target particle size would be with the fine generator.

References:

Agimelen, O., Svoboda, V., Ahmed, B., Cardona, J., Dziewierz, J., Brown, C., . . . Sefcik, J. (2018). Multi-sensor inline measurements of crystal size and shape distributions during high shear wet milling of crystal slurries. *Advanced Powder Technology, 29*(12), 2987-2995. doi:10.1016/j.apt.2018.09.003

Ahmed, B., Brown, C., McGlone, T., Bowering, D., Sefcik, J., & Florence, A. (2019). Engineering of acetaminophen particle attributes using a wet milling crystallisation platform. *International Journal of Pharmaceutics, 554*, 201-211. doi:10.1016/j.ijpharm.2018.10.073

Austin, L., Shoji, K., Bhatia, V., Jindal, V., & Savage, K. (1976). Some Results on the Description of Size Reduction as a Rate Process in Various Mills. *Industrial & Engineering Chemistry Process Design and Development*(1), 187. doi:10.1021/i260057a032

Bosetti, L., & Mazzotti, M. (2020). Population Balance Modeling of Growth and Secondary Nucleation by Attrition and Ripening. *Crystal Growth & Design, 20*(1), 307-319. doi:10.1021/acs.cgd.9b01240

Brown, C. J., McGlone, T., Yerdelen, S., Srirambhatla, V., Mabbott, F., Gurung, R., . . . Florence, A. J. Y. (2018). Enabling precision manufacturing of active pharmaceutical ingredients: workflow for seeded cooling continuous crystallisations. *Molecular Systems Design & Engineering*(3), 518. doi:10.1039/c7me00096k m3 - 10.1039/c7me00096k

Karabelas, A. (1976). Particle attrition in shear flow of concentrated slurries. *Aiche Journal, 22*(4), 765-771. doi:10.1002/aic.690220421

Szilagyi, B., & Nagy, Z. (2019). Model-based analysis and quality-by-design framework for high aspect ratio crystals in crystallizer-wet mill systems using GPU acceleration enabled optimization. *Computers & Chemical Engineering, 126*, 421-433. doi:10.1016/j.compchemeng.2019.04.025

138e Process Development and Characterization of an Acutely Hazardous Reaction through Data Rich Experimentation

Taylor Behre, Ben W. H. Turnbull, Erin McCarthy, Nelo R. Rivera, Alexei Kalinin, Brittany Armstrong, Ji Qi, Andrew Solovyov, Jonathan McMullen

Nemtabrutinib is an oral small molecule reversible inhibitor of Bruton's tyrosine kinase for the treatment of chronic lymphocytic leukemia (CLL) and Non-Hodgkin's Lymphoma. A key reaction engineering challenge in the synthesis of the isolated penultimate involved a reduction via an acutely toxic gaseous reagent. This presentation will focus on the process development and scale-up of this reaction to control a key impurity that has low rejection downstream in the active pharmaceutical ingredient (API). The multi-phasic reaction required pressurized operation to limit the formation of the reaction impurity. This and the hazardous nature of the reaction created challenges in obtaining detailed process understanding around the highly non-linear reaction profile. Data rich experimentation (DRE) techniques were leveraged to avoid altering the reactor headspace during sampling and to avoid exposure to hazards. The multi-stage reaction was characterized through these methods to understand the various factors impacting the reaction and the impurity generation rates. Through this enhanced process understanding, a robust process was implemented at manufacturing scale to consistently produce over 200 kilograms of quality drug substance intermediate across multiple batches.

138g A Surrogate-Based Multi-Objective Optimization with Adaptive Sampling for Advanced Pharmaceutical Manufacturing

Yingjie Chen, Marianthi Ierapetritou

Incentivized by increasing market competition, pressure to maximize profit while maintaining product quality, and initiatives from regulatory entities to develop agile, flexible, and robust manufacturing lines, pharmaceutical manufacturing processes are shifting from batch to continuous operations[1-3]. With extensive research efforts, progress has been made in modeling solid-based continuous manufacturing processes[3,4]. Based on the developed models, mathematical optimization can be conducted to identify the best operating conditions to achieve different objectives, such as minimizing cost and energy consumption while maintaining product quality. The optimization results from *in silico* simulations can guide the experimental team in searching for optimal conditions, saving valuable time in the process development[5]. Some optimization case studies in continuous pharmaceutical manufacturing include the optimization of a crystallization process[6-9], a continuous direct compaction process[3], and an end-to-end system consisting of upstream crystallization of active pharmaceutical ingredients and downstream tableting process[7,10].

As simulations become more complex to include more accurate representation of process dynamics, the computational complexity also increases[11]. Thus it can be inefficient to solve the optimization problems with traditional optimization approaches and may lead to suboptimal solutions within limited time[11]. To address such difficulty for computationally intense models, surrogate-based optimization strategies have been proposed as a promising alternative. A surrogate model is built to approximate complicated models, and it is iteratively updated with using an adaptive sampling strategy that searches for new promising points based on specific infill criteria. The workflow is repeated until a user-defined stopping criterion is met, and the final surrogate model with low approximation error is used to identify the near-optimal solution[12-17]. Previous work has applied this approach to the continuous direct compaction process by using a weighted expected improvement (EI) function as an infill criterion to guide the search for new sample points toward feasible regions with low objective values[11,18]. A modified EI on the objective follows this step to search for global optimum within the identified feasible region[11]. Although the work demonstrates high accuracy in obtaining both feasible region boundary and the optimum, it is only applied to single-objective problems.

In this work, an updated framework of surrogate-based, feasibility-driven, multi-objective optimization with adaptive sampling is proposed to consider multiple objectives. Each objective function in the multi-objective problem is approximated using a surrogate. The constraints are grouped into a feasibility function based on maximum constraint violation and substituted with another surrogate model. For the infill criteria, both the centroid method and the expected hypervolume improvement (EHVI) method are implemented. The centroid method computes EI based on the first moment of the joint probability density function of the objectives. The EHVI method seeks Pareto solutions based on the difference of hypervolumes between the current and the next sample set. Following the identification of the Pareto front, a goal programming approach is implemented to provide guidance on the best solution. To demonstrate the effectiveness of the proposed framework, an example benchmark problem, and a case study of continuous pharmaceutical manufacturing process via the wet granulation route are presented. Sampling requirement, computational time, and solution accuracy resulting from the framework are compared against the optimization results obtained using surrogates without the adaptive sampling strategy. The framework is shown to have better performance in obtaining more accurate results with smaller sampling requirements. The proposed approach can thus become an integrated part of the Pharma 4.0 framework and effectively used by industry to investigate multiple competing objectives under quality constraints, allowing for a better decision-making strategy.

References

1. Ierapetritou M, Muzzio F, Reklaitis G. Perspectives on the continuous manufacturing of powder-based pharmaceutical processes. *AIChE Journal.* 2016;62(6):1846-1862.

2. O'Connor TF, Yu LX, Lee SL. Emerging technology: A key enabler for modernizing pharmaceutical manufacturing and advancing product quality. *Int J Pharm.* 2016;509(1-2):492-498.

3. Wang Z, Escotet-Espinoza MS, Ierapetritou M. Process analysis and optimization of continuous pharmaceutical manufacturing using flowsheet models. *Comput Chem Eng.* 2017;107:77-91.

4. Metta N, Ghijs M, Schäfer E, et al. Dynamic flowsheet model development and sensitivity analysis of a continuous pharmaceutical tablet manufacturing process using the wet granulation route. *Processes.* 2019;7(4):234.

5. Rogers AJ, Inamdar C, Ierapetritou MG. An Integrated Approach to Simulation of Pharmaceutical Processes for Solid Drug Manufacture. *Industrial andamp; Engineering Chemistry Research.* 2014;53(13):5128-5147.

6. Sen M, Rogers A, Singh R, et al. Flowsheet optimization of an integrated continuous purification-processing pharmaceutical manufacturing operation. *Chemical Engineering Science.* 2013;102:56-66.

7. Patrascu M, Barton PI. Optimal Dynamic Continuous Manufacturing of Pharmaceuticals with Recycle. *Industrial andamp; Engineering Chemistry Research.* 2019;58(30):13423-13436.

8. Su Q, Benyahia B, Nagy ZK, Rielly CD. Mathematical Modeling, Design, and Optimization of a Multisegment Multiaddition Plug-Flow Crystallizer for Antisolvent Crystallizations. *Organic Process Research andamp; Development.* 2015;19(12):1859-1870.

9. Zhao Y, Kamaraju VK, Hou G, Power G, Donnellan P, Glennon B. Kinetic identification and experimental validation of continuous plug flow crystallisation. *Chemical Engineering Science.* 2015;133(Complete):106-115.

10. Patrascu M, Barton PI. Optimal campaigns in end-to-end continuous pharmaceuticals manufacturing. Part 2: Dynamic optimization. *Chemical Engineering and Processing - Process Intensification.* 2018;125:124-132.

11. Wang Z, Escotet-Espinoza MS, Singh R, Ierapetritou M. Surrogate-based Optimization for Pharmaceutical Manufacturing Processes. In: Espuña A, Graells M, Puigjaner L, eds. *Computer Aided Chemical Engineering.* Vol 40. Elsevier; 2017:2797-2802.

12. Jones DR, Schonlau M, Welch WJ. Efficient Global Optimization of Expensive Black-Box Functions. *Journal of Global Optimization.* 1998;13(4):455-492.

13. Bhosekar A, Ierapetritou M. Advances in surrogate based modeling, feasibility analysis, and

optimization: A review. *Comput Chem Eng.* 2018;108:250-267.

14. Boukouvala F, Ierapetritou MG. Derivative-free optimization for expensive constrained problems using a novel expected improvement objective function. *AIChE Journal.* 2014;60(7):2462-2474.

15. Wang Z, Ierapetritou M. Constrained optimization of black-box stochastic systems using a novel feasibility enhanced Kriging-based method. *Comput Chem Eng.* 2018;118:210-223.

16. Wang Z, Ierapetritou M. Surrogate-based feasibility analysis for black-box stochastic simulations with heteroscedastic noise. *Journal of Global Optimization.* 2018.

17. Boukouvala F, Ierapetritou M. Surrogate-Based Optimization of Expensive Flowsheet Modeling for Continuous Pharmaceutical Manufacturing. *Journal of Pharmaceutical Innovation.* 2013;8(2):131-145.

18. Metta N, Ramachandran R, Ierapetritou M. A novel adaptive sampling based methodology for feasible region identification of compute intensive models using artificial neural network. *AIChE Journal.* 2020;67(2).

81c Understanding of Heat Transfer in Freeze-Dryer during Primary and Secondary Drying Stages: Considerations of Vial's Thermal Properties

Kyu Yoon, Vivek Narsimhan

Freeze drying is a widely used process that improves the stability of pharmaceutical products. This process is time and energy intensive, and thus there is a strong need to understand and improve its heat transfer. Freeze drying consists of two stages: (a) primary drying (sublimation of ice), and (b) secondary drying (desorption of bound water). For primary drying, the heat transfer resistance between a shelf and a vial is inferred through changes in mass, and it is often assumed that this resistance remains the same during secondary drying. However, our recent study demonstrated that the heat transfer coefficient during secondary drying is much different than the measured value during primary drying due to changes in moisture content in the chamber. Additionally, unlike primary drying, the thermal properties of the vial play a much larger role in the temperature profile during secondary drying. To deeply investigate these observations, we performed careful measurements of the heat transfer in freeze dryer during primary and secondary drying using a lab-scale lyophilizer.

The purpose of this study is to determine the impact of different vials and their thermal properties on the vial heat transfer throughout the primary and secondary drying stages. We lyophilized different sugars (e.g., sucrose and mannitol) under a wide range of operating conditions in order to determine their influence on the energy transfer during secondary drying. Gravimetric and heat flow analysis methods were used to calculate the vial heat transfer coefficient K_V for multiple vial systems in a lab-scale freeze-dryer. Secondly, an energy budget analysis was used to determine the influence of the thermal mass of vials on the heat transfer characteristics of the secondary drying process.

Our findings revealed distinct heat transfer characteristics during the primary and secondary drying processes, as well as changes in the heat transfer coefficient for various vial compositions. Correlation parameters between heat transfer coefficient and chamber pressure are estimated using the data collected from various vials. Heat transfer characteristics explored in this study have great potential for minimizing the uncertainties in secondary drying modeling. It is expected that our results will aid in understanding and optimizing secondary drying for the lyophilized product.

81d Prediction of Tablet Dissolution from Real-Time Optical Coherence Tomography Data

Elisabeth Fink, Stephan Sacher, Varun Kushwah, Maxwell Korang-Yeboah, Xin Feng, Huiquan Wu, Johannes G. Khinast

The thickness and other quality aspects of pharmaceutical coatings are major impact factors for one of the most important quality attributes of a tablet, the dissolution profile. These quality attributes are conventionally determined by off-line measurement of weight or diameter gain. Optical Coherence Tomography (OCT) has attracted attention for at-line investigation and in-line monitoring of pharmaceutical film coatings [1]. This novel technology allows fast and non-destructive determination of thickness and other quality aspects of a representative number of coated tablets directly from the coating process. Implementation of OCT for an industrial scale pan coating process with capabilities for automated layer detection, classification (i.e. automated detection of tablets and layers), and thickness calculation in real time have been demonstrated previously [2].

This enables a systematic investigation of the link between coating quality attributes (e.g. coating thickness) measured in real-time and the target product quality attributes (e.g. dissolution). The goal of this study is to use the coating attributes measured in-line via OCT to develop a correlation model for the prediction of the expected dissolution profile, if a sample was taken at any point during the process. The talk will present the general workflow regarding how to develop a correlation model. The general workflow is applicable to obtain models for different formulations of tablets and pellets, which are coated via pan coating and fluidized bed coating, respectively. It will be demonstrated, how these models can be used to determine an ideal trajectory for the observed real-time values obtained via OCT, i.e. a golden batch trajectory, hence enhancing process robustness and enabling in-line quality control. The approach was developed and implemented in four representative use cases. The technology and method were transferred to different industrial sites such that their applicability across different industrial sites can be evaluated. The feasibility of this approach to contribute towards robust real-time release testing will be discussed.

[1] H. Lin, Z. Zhang, D. Markl et al., A Review of the Applications of OCT for Analysing Pharmaceutical Film Coatings. Appl. Sci. 2018; 8: 2700.

[2] S. Sacher, P. Wahl, M. Weißensteiner et al. Shedding Light on Coatings: Real-time Monitoring of Coating Quality at Industrial Scale. Int. J. Pharm. 566 (2019) 57–66.

81e Imaging and Analysis of Lyophilized Structures By Micro-Computed Tomography

Isaac Wheeler, Vivek Narsimhan, Alina Alexeenko

Lyophilization, better known as freeze drying, is a key technology for manufacturing shelf-stable pharmaceuticals such as injectable antibiotics, monoclonal antibodies, antibody-drug conjugates and vaccines. The lyophilization process can be divided into three main phases: freezing, primary drying, and secondary drying. Optimization of the freeze drying process requires a robust understanding of each stage of the process. For instance, the microstructure of the drug product depends on both the drug formulation and freezing process. The product microstructures in turn determine key parameters for both the primary and secondary drying stages, as well as the final appearance and properties of the dried drug product.

Microcomputed tomography (micro-CT) scanning is a powerful imaging technique, used to resolve 3D volumetric structures at resolutions as fine as 0.5μm. This is sufficient resolution to shed light on the microstructures that result from the lyophilization process. However, only a handful of studies have been published which used micro-CT scanning to image lyophilized structures, and few quantitative descriptions of the observed microstructures are given. In this work, I will present a more thorough geometric characterization of lyophilized mannitol and sucrose, including features such as porosity, tortuosity, and pore size distribution. This analysis will be carried out by developing a robust image processing procedure, using open source tools written in the Julia programming language, for the manipulation and analysis of micro-CT scans. The open source nature of the computational tools will make these tools more accessible to other researchers as well.

In future work, geometric parameters of the microstructure may be related to controllable parameters (e.g. freezing process, drug formulation) and useful macroscale parameters (e.g. mass transfer resistance, reconstitution time), allowing effective estimation of macroscale parameters without the need for lengthy process experiments.

81f Minimal Purification of Recombinant Proteins for Early Developability Assessment

Sergio Rodriguez Aponte, J. Christopher Love

Characterization of protein therapeutics and subunit vaccine candidates is a critical step in the protein development phase. Selection of lead candidates can be done by performing a developability assessment, which includes applying a battery of biophysical assays and selecting for ideal properties in accordance with the target product profile. To perform a developability assessment, however, several milligrams of purified protein is required, often delaying this assessment to the later stages of the development phase or obligating the use of affinity tags. Here we describe the use minimal purification process for recombinant proteins secreted by *Komagataella phaffii*. A buffer exchange and filtration with a Q-membrane filter proved to be sufficient to remove key supernatant impurities like host-cell proteins and DNA. This purification process yielded acceptable purity and 1-2 milligrams to characterize an engineered SARS-CoV-2 RBD by several widely used methods like mass spectrometry and differential scanning calorimetry. Furthermore, we used this method to perform a biophysical and functional comparison of SARS-CoV-2 RBD variants. Lastly, we demonstrate broader applicability to three other protein classes: a single-domain antibody, an NRRV protein subunit, and a human growth hormone. Implementation of a minimal purification strategy could streamline feedback loops between discovery and development teams in order to design more developable protein candidates.

81g In-Line Skin Thickness Evaluation on Co-Extruded Vaginal Rings By Means of Optical Coherence Tomography

Matthias Wolfgang, Ioannis Koutsamanis, Martin Spörk

Hot-melt extrusion (HME) has been extensively used for the manufacturing of a wide array of drug delivery systems such as granules, tablets and implants. One specific type of HME process enables the manufacturing of multilayer systems such as core-membrane or 'reservoir-type' formulations. Such systems consist of a core containing the active pharmaceutical ingredient(s) (APIs) in a polymeric carrier and a drug-free membrane, which controls the permeation and release of the drug in the surrounding environment. This design results in a constant, zero-order release over long times, which, in turn, translates into high patient compliance. For a given system, i.e. API loading and types of polymers used, the main contribution to the drug release falls on the membrane thickness and the membrane homogeneity. Commonly, the membrane thickness of co-extrudates is evaluated offline via optical microscopy or μ-CT imaging of single cross-sectional cuts. However, both methodologies are associated with several drawbacks: in both cases, the assessment is time-consuming and not destruction free due to the required cuts or the exposure to ionizing radiation. Moreover, the results might not be representative of the extrusion process, since only fractions of a sample are investigated from a whole batch and manual cutting might induce additional changes in the membrane due to mechanical stress. Consequently, a fast, non-destructive and reproducible method to evaluate the membrane thickness of co-extruded reservoir systems is of great interest.

Our current study intends to address this gap by presenting an in-line capable, non-invasive methodology based on optical coherence tomography (OCT) to evaluate the membrane characteristics of ethylene-vinyl acetate (EVA) based co-extrudates, aiming at quicker but simultaneously more representative results. The presented approach opens the way to demonstrate the capabilities of OCT as an inline quality assurance tool for the manufacturing of the next generation of co-extruded formulations. OCT as a technology is well-known and established in the fields of medicine, art conservation, and non-destructive testing. Furthermore, RCPE has pioneered its use in pharmaceutical applications, specifically for real-time monitoring the coating thickness of solid dosage forms such as tablets and pellets. While measuring coating thickness and its uniformity is the basic functionality of pharmaceutical OCT systems, the potential of the approach is far greater. OCT is capable of detecting abrupt changes in material refractive index and scattering properties. Hence, it can visualize internal structures and morphology e.g., delamination and excipient agglomerates with sufficient size within the membrane structure.

In our talk, we will present an approach to use OCT as an in-line tool for co-extrusion processes. We will compare the results obtained by means of OCT and compare them to state of the art microscopic investigation of cross-sectional cuts. Aspects discussed will be advantages and disadvantages of both methods and obtained results. One considerable advantage of OCT is the real time capability, reporting instantaneous results. As such, the influence of the strand speed during extrusion was investigated as well in a range of 65 – 130 mm/sec. Investigations on extruded strands revealed one additional merit of using OCT, as details on the membrane joint, where the co-extruder provides the polymer melt generated some defects became obvious and quantifiable. In summary, we believe that the use of OCT as a process analytical technology in HME, and in particular co-extrusion processes, has the potential to act as a game changer. Reasons to believe so are the fast and destruction free measurements, the high resolution and wealth of information obtainable.

337a Optimization of Continuous Spin Freezing in Single Vial Unit By Implementing Computational Fluid Dynamics Modeling and Simulation

Isar Charmchi, Gust Nuytten, Ashish Kumar, Thomas De Beer

Freeze-drying is a common method for increasing the stability and shelf life of heat-sensitive and water-labile (bio)pharmaceuticals. The pharmaceutical sector is currently transitioning from batch to continuous manufacturing. According to the recent FDA guideline regarding continuous manufacturing of drug substances and drug products (Q13), maintaining a level of control in CM necessitates an understanding of process dynamics. Three consecutive phases are involved in the pharmaceutical freeze-drying process: freezing, primary and secondary drying. Using cold sterile gas and rapid rotation of glass vials along their longitudinal axis, the temperature of the aqueous solution is lowered during the spin freezing stage until ice starts to nucleate, followed by ice growth. The gas flow and temperature are controlled during the spin freezing process to manage the ice crystal size distribution. During the primary drying, the chamber pressure is lowered considerably below the vapor pressure of the ice. Further, an infrared (IR) heater supplies energy for sublimation during the primary drying process. Desorption of the unfrozen water takes place during secondary drying. Being the first step in the dehydration process, the freezing phase in the lyophilization process is crucial as it determines the ice shape, the pore size of the final dried product and hence the drying conditions. To understand the dynamics during the spin freezing step as part of a recently developed continuous freeze-drying process, for the first time, a detailed model based on computational fluid dynamics (CFD) is developed (CFX v.19.2, Ansys Inc.) for the single vial unit (SVU) freeze dryer. The CFD simulation provides the vial temperature and a local gas flow pattern inside the cooling chamber of the investigated system.

The rotational speed of the vial during spin freezing and the velocity of the cooling gas which both - next to the gas temperature - determine the cooling rate of the product varies from 2000 rpm to 5000 rpm and 2 to 20 m/s, respectively. Multiple reference frame (MRF) and moving mesh modeling methodologies are being studied for simulating the gas flow inside the chamber. To simulate a steady-state flow, MRF is employed. The MRF simulation results are utilized as the initial condition for the moving mesh technique to model flow in the transient state. Further, the heat equation is solved to simulate the vial temperature. The temperature along the vial is measured using the Thermal IR camera (Fig. 1). The experimental data support the temperature profile obtained through simulation. The validity of the gas flow behavior in the chamber was analyzed using the tracer simulation. Smoke test is used to confirm the tracer simulation. The smoke flow in the chamber is recorded using a high speed camera. The findings of the simulation and the experiments are in good agreement.

Finally, the design will be further optimized using the verified and validated CFD simulation. To optimize the cooling rate, several geometric factors such as nozzle size and placement, gas outlet location, and dimension will be studied. Thus, detailed mechanistic modeling improves understanding of process dynamics and aids in the optimization of the spin freezing set up.

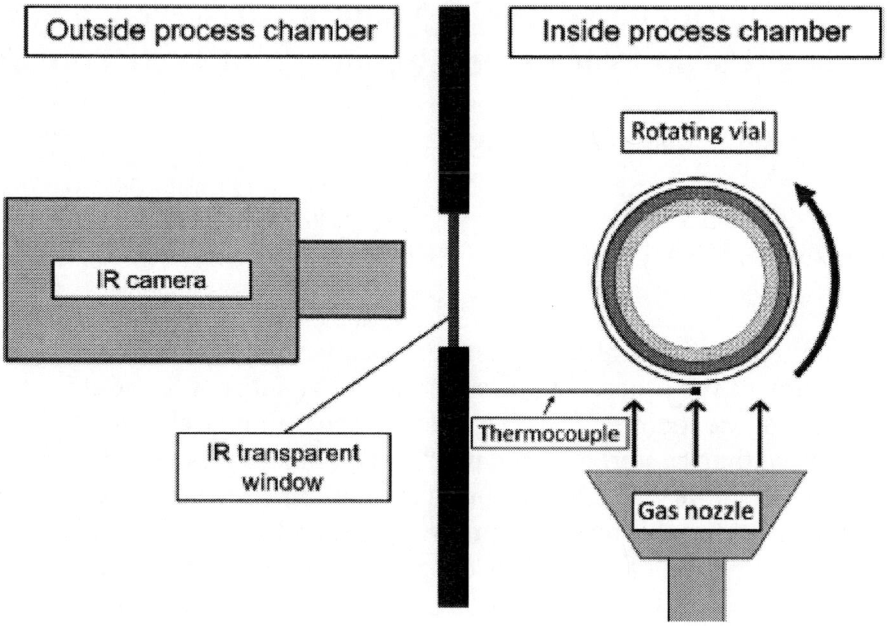
Outside process chamber
Inside process chamber
Rotating vial
IR camera
Thermocouple
IR transparent
window
Gas nozzle

337b A System Model for a Dry-Granulated Tablet Manufactured from a Spray Dried Dispersion Intermediate: Use in Process Development and Tech Transfer

Salvador Salvador Garcia, Mehuli Kulkarni, James E. Miesle

This presentation addresses the development of an end-to-end model for an oral dosage form manufactured via dry granulation and utilizes a drug product intermediate manufactured by spray dried dispersion.

The model is built to predict key quality attributes of the tablet (such as dissolution and breaking force needed to crush the tablet) as a function of the processing parameters of the entire train of unit operations and the physical properties of the incoming SDD material. The multivariable relationships across material, process and product are thus captured by the model. The model was used to derive a multi-dimensional design space using flexibility analysis and was also used to de-risk unexpected variability in the properties of the SDD material. The presentation will address mathematical aspects of the model, such as the over-parameterization of the well-accepted Johansson model, how to solve these and extend the model to be a function of the physical properties of the raw materials. Practicalities regarding the implementation and use will also be discussed.

337c Why Do We Dwell on Dwell Time? a Compaction Simulator Study Investigating Tableting Attributes

MayLin Howard, Gerard R. Klinzing

Compaction of powders and granules into tablets with robust mechanical strength and absence of structural defects remains a significant formulation challenge in the pharmaceutical industry. Two compression process parameters, dwell time and press speed (ex. punch velocity), are discussed as principal variables that directly contribute to the tableting process performance and properties, such as density, tensile strength, and defects. Specifically, extended dwell time has been correlated to reduce the chance of capping and lamination, increase tablet density, and increase tensile strength. The proposed mechanism is longer dwell times allow for stress relaxation and release of entrapped air, making critical defects less likely. On the other hand, reduced punch velocity has been attributed to decreased air entrapment and thus lower incidence of tablet defects/cracking. Prior work is limited, and many studies do not independently control dwell time and punch velocity due to equipment limitations. In this study, we systematically investigate the independent influence of dwell time and punch velocity using a hydraulic compaction simulator. In anticipation of the impact of dwell time and punch velocity being material-dependent, we used a range of material classes such as: ductile materials (ex. microcrystalline cellulose), elastic materials (ex. polyethylene oxide), and brittle materials (ex. dicalcium phosphate). Measurements included tensile strength, tablet density, tablet ejection force, stress relaxation, and imaging of critical tablet defects like capping and lamination through visual inspection and x-ray microcomputed tomography. By understanding the individual effects of dwell time and punch velocity on powder compaction, researchers can avoid unexpected observations on high-speed rotary presses. Decoupling their effects may encourage appropriate reporting of the crucial process parameter(s) for each material and enable appropriate risk responses for tablet defects to improve mechanical property performance. Moreover, understanding of the mechanism responsible for tablet properties, compaction velocity or dwell time, will inform the development of more robust production equipment.

337d Uncertainty Propagation for Probabilistic Prediction in Partial Least Squares Using Bootstrap Methods

James Odgers, Chrysoula Kappatou, Ruth Misener, Salvador Salvador Garcia, Sarah Filippi

Partial Least Squares (PLS) is an important modeling technique within pharmaceutical manufacturing that is able to both predict the output space and describe a significant portion of the variance of the input space, through projection on a latent space. This allows PLS to perform well in situations with large numbers of highly correlated input variables. Unfortunately, the latent space found by PLS is non-linearly dependant on the output variables of the training data [1] - a feature that vastly complicates any probabilistic analysis of the errors associated with the technique. This work presents a novel method, based on bootstrapping [2], for calculating a probabilistic output from PLS that automatically takes into consideration the aforementioned non-linearities in the parameter uncertainty.

The main result of this work is a novel probabilistic prediction from PLS, which we demonstrate can differ significantly from predictions that avoid dealing with the non-linear dependence of the predictions on the training data's observed outputs. The benefit of these probabilistic predictions are demonstrated in design space identification. First, we use simulated data to compare our method's performance to those from a perfect model. In addition, we investigate our method's performance on real world data and demonstrate its ability to make predictions that are a reasonable representation of the difference between the model prediction and the true observed data.

[1] Helland, I. S. On the structure of partial least squares regression. Communications in Statistics-Simulation and Computation 17, 581–607 (1988).

[2] Fushiki, T., Komaki, F. andamp; Aihara, K. Nonparametric bootstrap prediction. Bernoulli 11, 293–307 (2005).

337e The Use of Integrated *in-Vitro-in-Silico* frameworks Toward De-Risking Low-Dose Dry Powder Inhalation Development

Joana Pinto, Snezana Radivojev, Valerie Reinisch, Eleonore Fröhlich, Amrit Paudel

Introduction

Understanding the impact that different development decisions have on the final product's performance is vital when attempting to de-risk the drug development process. For this, it is critical to apply *in-vitro* and *in-silico* tools to predict the *in-vivo* performance of the developed drug. Likewise, this abstract reports a series of relevant case studies where the integrated use of distinct *in-vitro* and *in-silico* approaches has been found useful to support the development of low-dose dry powder inhalers (DPI) (Figure 1). The proposed integrated framework was used to guide the selection of bio-relevant solubility (study 1) and dissolution methodologies (study 2), help guide formulation (study 3) and process selection (study 4), and support the prediction that DPI misuse could have on therapeutic outcomes (study 5).

Materials and Methods

Different inhaled drug and carrier materials and their formulations, both in-house engineered, and commercial products were tested. For the in-house formulations, salbutamol sulphate (SS), budesonide (BUD), and formoterol fumarate (FF) were used as model active pharmaceutical ingredients (APIs) and lactose (LAC) and mannitol (MAN) as carriers. Two commercial products comprised of BUD were also tested to evaluate their in-use stability.

Solubility determination

The solubility of the different APIs was determined via the shake-flask method using distinct simulated lung fluids (SLFs) containing PBS (pH 7.4) alone or with varying combinations of albumin (protein) and lipids. An excess amount of the drugs was added to 5 mL of the media and incubated at 37°C at a shaking speed of 100 rpm for 24 h.

Adhesive mixing and blend uniformity

Adhesive mixing yielded DPI formulations containing blends of the APIs and carrier excipient particles. Mixing was performed at low and medium shear. For the low shear rate blending a Turbula blender TC2 (Willy A. Bachofen Maschinenfabrik, Muttenz, Switzerland) was applied for 90 min at 62 rpm. At medium shear, the blends were mixed in a laboratory-scale resonant acoustic mixer (LabRAM I ResonantAcoustic® Mixer, Resodyn^{TM} Acoustic Mixers, USA) for 90 s at two acceleration levels (i.e., 30 and 60 g). The blend uniformity was evaluated by measuring the content of API in 10 samples taken from different positions of the powder bed, and considered uniform at a relative standard deviation < 10%.

Evaluation of the in-vitro aerodynamic performance

The *in-vitro* aerodynamic performance was evaluated via next generation impactor (NGI, Copley, UK) studies. The NGI experiments were carried out with different airflows and using different inhalers. For each scenario, it was ensured that 4 L of air was drawn via the mouthpiece over the inhaler.

In-vitro evaluation of the dissolution performance

For the dissolution, the fine particle fraction was first collected using a fast screening impactor (FSI, Copley, UK). The deposited particles were ''sandwiched'' using a polycarbonate (PC) filter placed in a crystallization dish with 30 mL of medium. In sequence, the dish was put into an orbital shaker at 37°C and 60 rpm, and the dissolution was performed for 4 h, taking sample aliquots at relevant time points. An SLF containing lipids was used as the medium, and sink conditions were maintained, with respect to the solubility of the drugs.

Modeling of the in-vivo deposition

The deposition of the APIs was predicted in the Multiple-Path Particle Dosimetry v2.11 software (MPPD, Applied Research Associates, Inc). The NGI deposition profiles were used to calculate the mean mass aerodynamic geometric (MMAD) and the geometric standard deviation (GSD). In combination with spirometry profiles of the devices, the MMAD and GSD were applied to calculate the deposition of the drug in the relevant populations.

In-silico modeling of the dissolution performance

The dissolution model was developed using SimBiology® (MathWorks Inc, version R2020a). A four compartmental model describing the mass of deposited solid particles, their dissolution, permeation, and collection was built. The compartments were interconnected via the dissolution, permeation, and collection rates. The dissolution profile was predicted based on the *in-vitro* values obtained for API solubility and permeability and the setup details.

Physiological based pharmacokinetic modeling

Based on the available *in-vivo* data following their inhalation, the physiological-based pharmacokinetic (PBPK) models for the APIs were developed using GastroPlus® (Simulations Plus, USA) in its integrated PCAT (Pulmonary Compartmental Absorption and Transit^{TM}) route module. For this, the using physicochemical properties of the API and formulations were used as inputs.

Case study results

Study 1: Influence of diverse media on APIs' solubility and predicted in-vivo performance

Screening of the impact of different SLFs on the determination of the solubility values to be used as inputs to predict the *in-vivo* plasma concentration of BUD and salbutamol (SAL) showed different results depending on the nature of the molecule. For BUD, the SLF combining lipids and protein resulted in solubility values that, when used *in-silico*, yielded the most similar profiles to the *in-vivo* ones. In this medium, the solubility of BUD increased approximately five times compared to pure PBS. On the contrary, the solubility of SS remained unaffected among all investigated media; thus, no impact was observed in the predictions. This indicated that it is critical to characterize the solubility of inhaled APIs and carefully select the media for this when attempting to predict the *in-vivo* performance of DPIs.

Study 2: Evaluation of the use of different prediction approaches for the mean dissolution time

Successful *in-vitro-in-vivo* correlations (IVIVC) are dependent on the reliability of the methods used to accomplish this end. In this study, two approaches were evaluated concerning their ability to predict the mean dissolution time (MDT) and establish a level B IVIVC. The first approach used experimental values obtained *in-vitro* for the BUD solubility and permeability to model the dissolution *in-silico*. In the second approach, the MDT was deconvoluted directly from the dissolution profiles obtained *in-vitro*. Comparison of the *in-vivo* MDT of BUD to the one obtained from the *in-silico* model showed a prediction error of 19%. The error was notably higher (around 50%) when the MDT was directly determined from the *in-vitro* dissolution profile of BUD and compared to the *in-vivo* one. This showed that the solubility and permeability values used as input parameters for the *in-silico* model were adequate, however, the *in-vitro* dissolution methodology used was not. This indicated that further improvements were necessary to make the *in-vitro* dissolution set-up more bio-relevant.

Study 3: Carrier selection guidance through a PBPK framework

LAC grades presenting distinct properties were selected as model carriers and blended with a low dose of SS to be tested using a reservoir and a capsule-based device at 28, 60 and 100 L/min. The blend containing the highest fraction of excipient fines (< 10 μm) was the one that was less affected by the airflow and the one resulting in urine concentrations of SAL similar to Diskus. This showed that a certain percentage of fines (≈10%) is necessary to assure the adequate bioavailability of SS at distinct flow rates, including low ones when respiratory disease patients are unable to produce strong inhalation forces.

Study 4: Impact of adhesive blending process parameters on the predicted DPI deposition in-vivo

To evaluate the different adhesive blending process parameters' impact on the release of the dual combination of BUD and FF formulations, the APIs were blended with an inhalable grade of MAN at different accelerations using a resonant acoustic mixer. By simulation of the deposition *in-vivo*, it was shown that the blending acceleration had a more notable impact on the deposition pattern of BUD when the capsule-based device was used, while, for FF more distinct impact was observed when the reservoir based inhaler was used. This set of observations underlines how a balance between the specific APIs' particle behavior needs to be considered when defining the optimal process space for combination DPI products.

Study 5: Predicting the risk of humidity exposure on the performance of different DPI products

The use of drug-device combinations by patients is prone to errors; thus, it is essential to understand from an early stage how susceptible a formulation could be to misuse. In this study, the erroneous storage of two different commercial BUD DPIs under high humidity environments was explored. Storage of the product presenting a larger carrier particle size and a lower amount of fines resulted in less variation of the predicted $AUC_{0-\infty}$ values, indicating less susceptibility to high humidity storage conditions. The results of this study showed how the integration of PBPK modeling could support the late phase development of formulations less susceptible to patient misuse.

Conclusions

The series of case studies here presented, illustrate how an *in-vitro-in-silico* framework can be integrated from early to late phase development of low dose DPI products, de-risking development. Fit-for-purpose *in-vitro* analytical data can be rationally utilized *in-silico* at different stages of DPI product development to connect particle, formulation, and process parameters to the pharmaceutical performance.

References

Radivojev, G. Luschin-Ebengreuth, J.T. Pinto, P. Laggner, A. Cavecchi, N. Cesari, M. Cella, F. Melli, A. Paudel, and E. Fröhlich, Int. J. Pharm. **606**, 120893. DOI: 10.1016/j.ijpharm.2021.120893

J.T. Pinto, I. Cachola, J.F. Pinto, and A. Paudel, Pharmaceutics **13** (3), 297. DOI: 10.3390/pharmaceutics13030297

Radivojev, J.T. Pinto, E. Fröhlich, A. Paudel, J. Drug Deliv. Sci. Technol. **52**, 803-817 (2019). DOI: 10.1016/j.jddst.2019.05.047

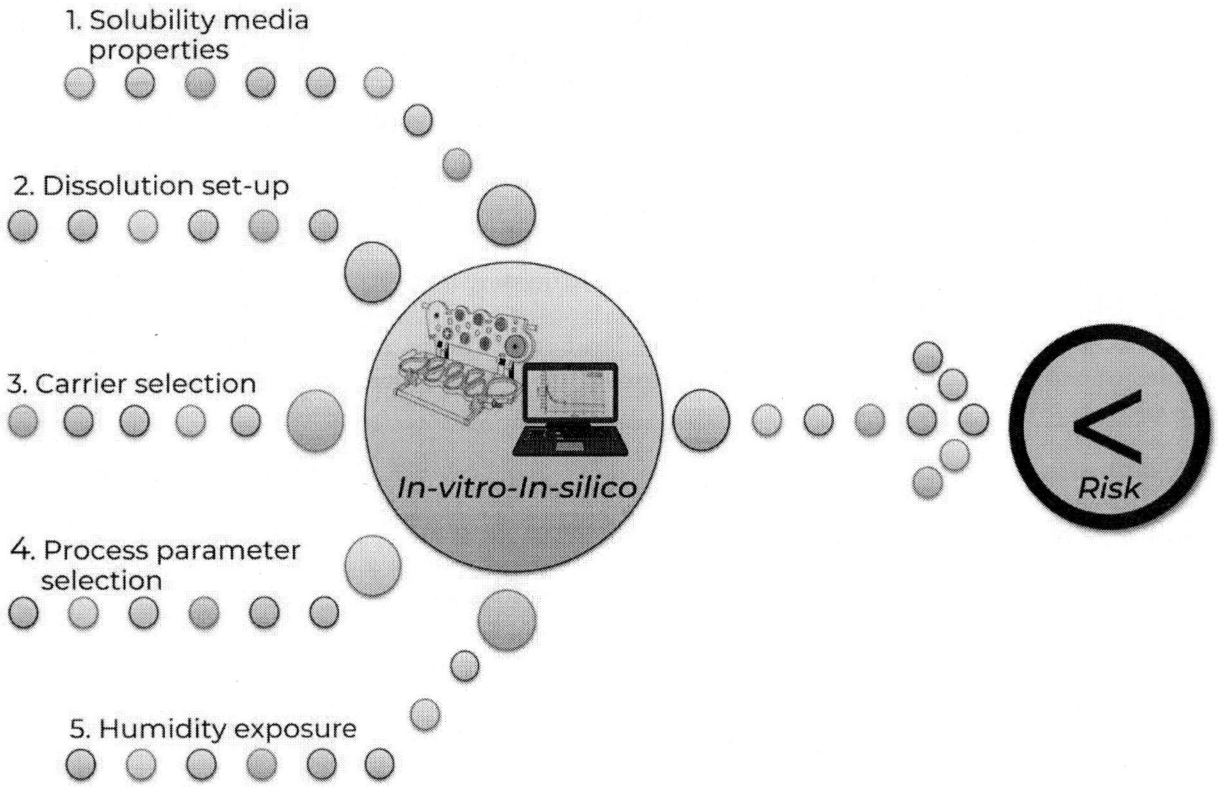

Figure 1: case studies where the use of *in-vitro-in-silico* approach was used to de-risk DPI development

337f Pharmaceutical HME Process Development: Understanding Non-Newtonian Fluid Flow

Josip Matic, Svetlana Marinova, Ralf Kühn, Markus Schmudde, Johannes G. Khinast

Introduction

Hot-melt extrusion (HME) is a continuous manufacturing process that uses co-rotating intermeshing twin-screw extruders (TSE). The process is primarily used to produce amorphous solid dispersions of poorly soluble active pharmaceutical ingredients (APIs) when used in pharmaceutical applications. In addition, it can also facilitate the development of products with crystalline API embedded into a polymer matrix and nano-pharmaceuticals. Being a continuous manufacturing technology, HME based drug development typically requires prohibitively high amounts of API (kilograms of premix vs. grams available in the early development stages) for successful formulation screening and early process development. The process setup itself is modular, allowing the process to be tailor-made for any formulation. This might, however, be problematic in choosing the appropriate and efficient design for an unknown formulation. It is an additional reason for the relatively high quantities of material usually needed for the screening. Effectively solving the multidisciplinary challenge of formulation development, early process screening, effective scale-up, and transfer to GMP production represents one of the critical challenges of the pharmaceutical industry and is especially challenging for novel technologies like HME.

Methodology

Addressing these challenges, our group has worked on developing *in silico* and experimental tools for more straightforward process development and scale-up. To this date, most of the process setup and scale-up activities are performed experimentally and empirically; one of the goals of our group is to create *in silico* tools for a rational, science-based process setup and scale-up while addressing other important aspects, such as API degradation and overall product quality [1], [2]. The fundamental idea behind process modeling is the breakdown and analysis of critical process aspects, among which the most prominent might be the analysis of flow patterns developed as a result of the rotation and geometry of the individual screw element pairs [1], [3]–[8]. Following our work showing the impact a separate kneading section along the screw configuration has on the product quality in [9], [10], our work focused on increasing the number of physical quantities we can quantify using *in silico* tools. For this purpose, we use the Smoothed Particle Hydrodynamics (SPH) simulation tool to cover non-Newtonian fluids and melt temperature prediction. Model validation and the characterization of different screw types were performed. The choses screw types are shown in Figure 1.

Results

Our work included computing the shear rate distribution, viscosity distribution, melt temperature distribution, local residence time distributions, pressure, and power characteristics of conveying, kneading, and mixing elements. The generated data is analyzed in shear rate, viscosity, local residence time, and melt temperature distribution. Every element type, cross-section geometry, operating condition, and formulation viscosity results in a different flow field. Understanding this flow field is critical in predicting the mixing ability of the selected screws. Moreover, different processing conditions also create a spectrum of residence time and melt temperature distributions that should be considered when assembling the screw configuration. The obtained data is used to develop reduced-order models that are more precise than simplified equations usually used in the 1D extrusion models. Better 1D extrusion models will help to predict the quality of the product, even before the first extrusion experiments, de-risk the development, and significantly reduce the time to market, cost, and waste produced in the development of novel drug forms.

Literature

[1] J. Matić, A. Witschnigg, M. Zagler, S. Eder, and J. Khinast, "A novel in silico scale-up approach for

hot melt extrusion processes," *Chem. Eng. Sci.*, vol. 204, pp. 257–269, Aug. 2019.

[2] J. Matić, A. Paudel, H. Bauer, R. A. L. Garcia, K. Biedrzycka, and J. G. Khinast, "Developing HME-Based Drug Products Using Emerging Science: a Fast-Track Roadmap from Concept to Clinical Batch," *AAPS PharmSciTech*, vol. 21, no. 5, p. 176, Jul. 2020.

[3] A. Eitzlmayr and J. G. Khinast, "Co-rotating twin-screw extruders: Detailed analysis of conveying elements based on smoothed particle hydrodynamics. Part 1: Hydrodynamics," *Chem. Eng. Sci.*, vol. 134, pp. 861–879, Sep. 2015.

[4] A. Eitzlmayr and J. G. Khinast, "Co-rotating twin-screw extruders: Detailed analysis of conveying elements based on smoothed particle hydrodynamics. Part 1: Hydrodynamics," *Chem. Eng. Sci.*, vol. 134, pp. 861–879, Sep. 2015.

[5] A. Eitzlmayr, J. Matić, and J. G. Khinast, "Analysis of flow and mixing in screw elements of corotating twin-screw extruders via SPH," *AIChE J.*, vol. 63, no. 6, pp. 2451–2463, Jun. 2017.

[6] A. Eitzlmayr *et al.*, "Experimental characterization and modeling of twin-screw extruder elements for pharmaceutical hot melt extrusion," *AIChE J.*, vol. 59, no. 11, pp. 4440–4450, Nov. 2013.

[7] A. Eitzlmayr *et al.*, "Mechanistic modeling of modular co-rotating twin-screw extruders," *Int. J. Pharm.*, vol. 474, no. 1–2, pp. 157–176, Oct. 2014.

[8] R. Baumgartner, J. Matić, S. Schrank, S. Laske, J. Khinast, and E. Roblegg, "NANEX: Process design and optimization," *Int. J. Pharm.*, vol. 506, no. 1–2, pp. 35–45, Jun. 2016.

[9] J. Matić *et al.*, "Towards predicting the product quality in hot-melt extrusion: Small scale extrusion," *Int. J. Pharm. X*, vol. 2, p. 100062, Dec. 2020.

[10] J. Matić *et al.*, "Towards predicting the product quality in hot-melt extrusion: Pilot plant scale extrusion," *Submitted.*

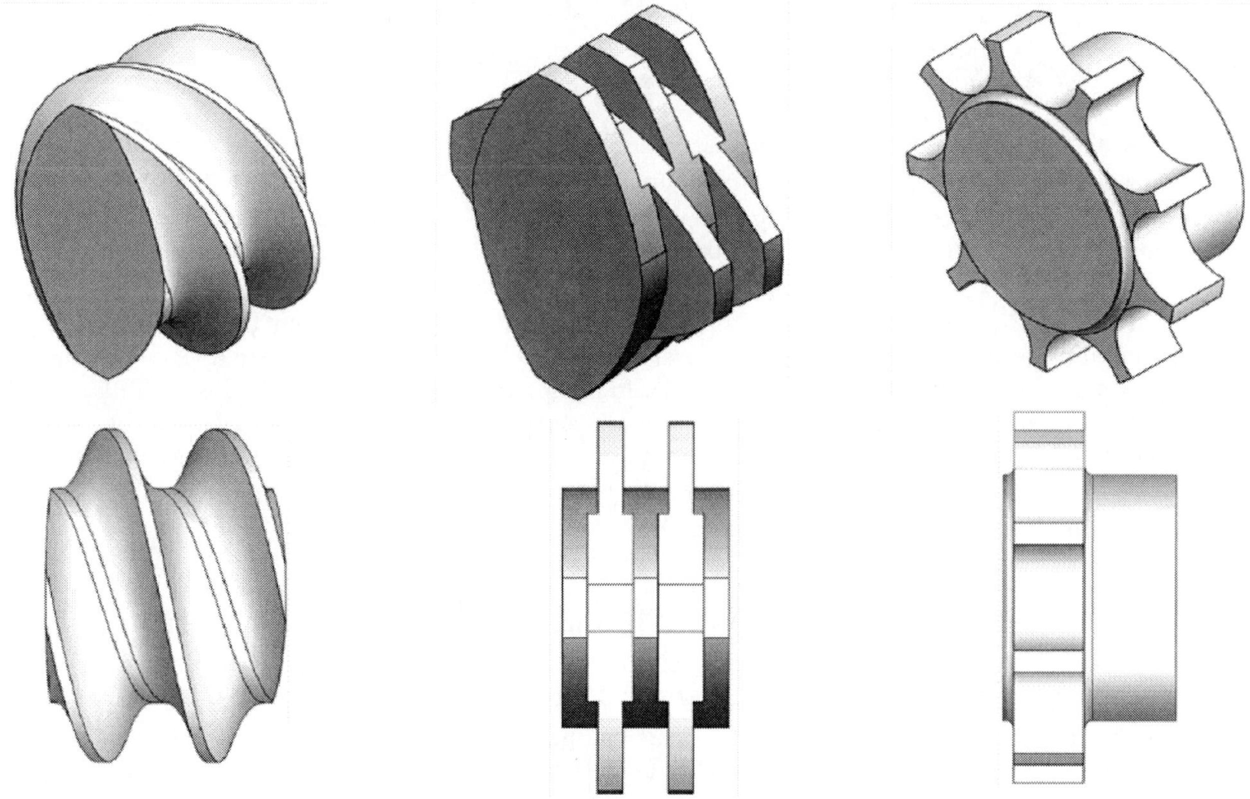

337g Integrated Formulation and Process Design for 3D Printing of Pharmaceuticals

Varun Sundarkumar, Zoltan Nagy, Gintaras Reklaitis

Pharmaceutical manufacturing has entered a phase of exciting transformation. The conventional batch production schemes that have been a fixture in the industry are giving way to more efficient, novel processes such as continuous manufacturing and 3D printing (3DP).[1] Partly driven by disruptions suffered in the pandemic, the industry has also realized the importance of developing distributed manufacturing networks for agile and flexible drug production. Recent studies have sought to combine these advances to build continuous mini-plants that can service an individual patient's drug requirement on demand.[2,3] For economic viability, these mini-plants need to be designed with a high degree of automated manufacturing capability. A major thrust in this effort has been to build integrated product and process design (PPD) frameworks to simultaneously design the product properties and manufacturing conditions based on prespecified performance characteristics desired from the product. The central idea in PPD problems is to solve the inverse estimation problem, i.e., given the functionality desired, design the product to match specifications and design the process to manufacture it.[4,5]

This study seeks to build an integrated PPD framework for a pharmaceutical 3D printing platform – Drop on demand (DoD) printing. DoD is a versatile inkjet 3DP that has been demonstrated to produce precise, personalized dosages for a variety of pharmaceutical formulations.[6,7] It also supports integration with upstream active ingredient synthesis steps and has been demonstrated to function as a drug product manufacturing unit of a continuous mini-plant.[2] Drug loading in the dosages produced by DoD need to be tightly controlled. Thus, the functionality desired in DoD is high printing consistency which is achieved when the drops generated have low variance in volume and have a low number of satellite drops. Printing consistency depends on both, formulation (product) properties like composition, viscosity, surface tension etc. and printing (process) conditions like drop ejection frequency, ejection pulse amplitude etc. For a novel formulation, the input (formulation and printing) conditions that give high printing consistency are conventionally determined experimentally. The PPD framework reported in this paper seeks to automate this process thereby engendering savings in time and material consumed.

To develop this framework however, a process model needs to be built to determine printing consistency for given input conditions. Theoretical studies modelling drop formation (occurrence of satellite drops, uniformity of drops generated etc.) in inkjets tend to focus on Newtonian fluid systems, however these cannot be directly applied to pharmaceutical suspensions where non-idealities such as polymeric carrier fluids and concentrated particle suspensions are frequently involved. Data driven approaches like machine learning can provide an alternate means to model the DoD printer.[8,9] There is a plethora of machine learning techniques available to model process operations, artificial neural networks are one of the most widely used among them.

Thus, as a prerequisite to building the PPD framework, this study seeks to develop a neural network based machine learning model for DoD. This model will predict a printing consistency metric for a given set of input conditions. To build this model, experimental data is generated for a variety of formulation and printing conditions. This model is then applied to build the PPD framework and answer the question: to achieve high printing consistency, what printing and formulation properties can be used. Data driven models give only a one-way relationship between the input variables and the output. This makes solving the inverse estimation problem challenging as specifying output (printing consistency) does not enable calculation of input conditions. To overcome this a two-stage approach is proposed: first, points in the input design space are sampled using a broad sampling scheme such as Latin hypercube. The model is evaluated at the sampled points, input conditions that give high printing consistency are then screened and used as input to the next stage. In the second stage, a series of heuristics are applied to select the most desirable solution from this set (figure 1).

In this paper we will demonstrate that this two-step approach can form an effective basis for a PPD framework for DoD that can automatically determine optimal formulation and printing conditions for manufacturing dosages of any given drug.

References:

1. *Innovations in Pharmaceutical Manufacturing on the Horizon.* National Academies Press; 2021. doi:10.17226/26009
2. Sundarkumar V, Nagy ZK, Reklaitis G v. Small-scale continuous drug product manufacturing using dropwise additive manufacturing and three phase settling for integration with upstream drug substance production. *Journal of Pharmaceutical Sciences*. Published online March 2022. doi:10.1016/j.xphs.2022.03.009
3. Adamo A, Beingessner RL, Behnam M, Chen J, Jamison TF, Jensen KF, Monbaliu JCM, Myerson AS, Revalor EM, Snead DR, Stelzer T, Weeranoppanant N, Wong SY, Zhang P. On-demand continuous-flow production of pharmaceuticals in a compact, reconfigurable system. *Science (1979)*. 2016;352(6281). doi:10.1126/science.aaf1337
4. Eden MR, Jørgensen SB, Gani R, El-Halwagi MM. A novel framework for simultaneous separation process and product design. *Chemical Engineering and Processing: Process Intensification*. 2004;43(5). doi:10.1016/j.cep.2003.03.002
5. Ng KM, Gani R. Chemical product design: Advances in and proposed directions for research and teaching. *Computers and Chemical Engineering*. 2019;126. doi:10.1016/j.compchemeng.2019.04.008
6. Içten E, Giridhar A, Taylor LS, Nagy ZK, Reklaitis G v. Dropwise additive manufacturing of pharmaceutical products for melt-based dosage forms. *Journal of Pharmaceutical Sciences*. 2015;104(5). doi:10.1002/jps.24367
7. Radcliffe AJ, Hilden JL, Nagy ZK, Reklaitis G v. Dropwise Additive Manufacturing of Pharmaceutical Products Using Particle Suspensions. *Journal of Pharmaceutical Sciences*. 2019;108(2). doi:10.1016/j.xphs.2018.09.030
8. Brishty FP, Urner R, Grau G. Machine learning based data driven inkjet printed electronics: jetting prediction for novel inks. *Flexible and Printed Electronics*. 2022;7(1):015009. doi:10.1088/2058-8585/ac5a39
9. Wang J, Chiu GTC. Data-driven drop formation modeling in nanoliter drop-on-demand inkjet printing. In: *ASME 2020 Dynamic Systems and Control Conference, DSCC 2020*. Vol 2. ; 2020. doi:10.1115/DSCC2020-3292

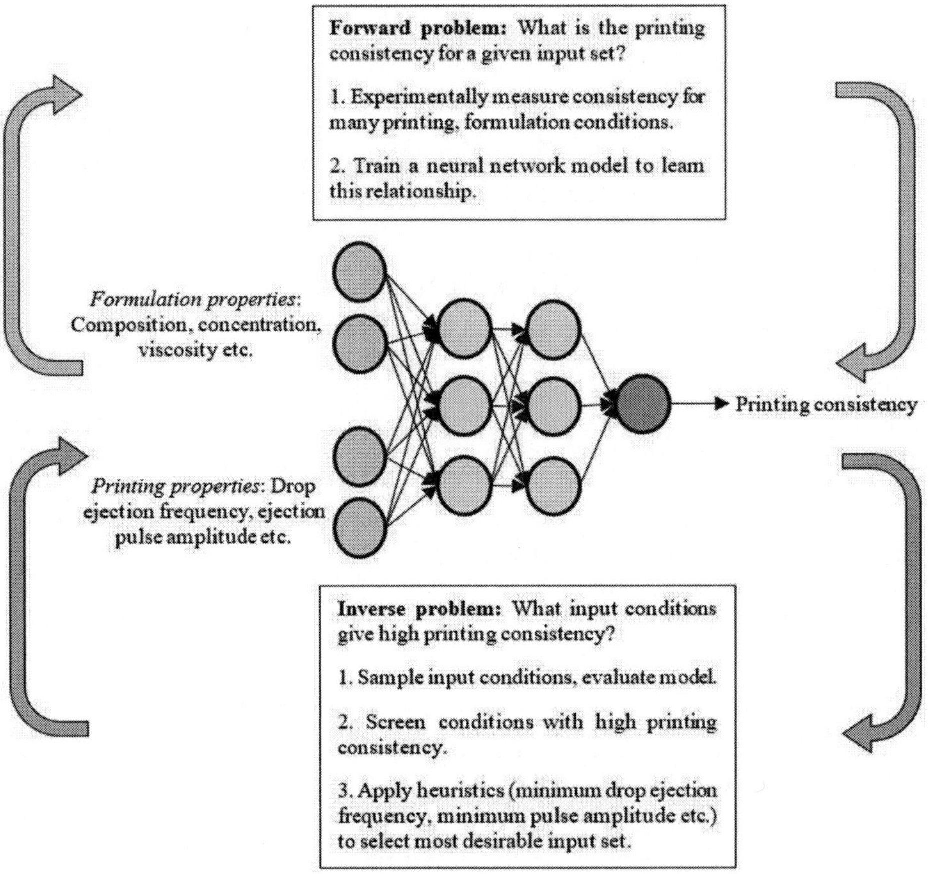

Figure 1. This figure summarizes the integrated PPD framework proposed in the study. First, a neural network model is built to predict printing consistency based on formulation and printing conditions. Then this model will be used to solve the inverse problem: to achieve high printing consistency, what input conditions should be used.

515a Development of a Flow Platform for Autonomous Model-Based Experimental Design

Christopher Hone, Sebastian Knoll, Clemens E. Jusner, Peter Sagmeister, Jason Williams, Martin Horn, C. Oliver Kappe

Currently, the development and optimization of flow processes is seen as a resource intensive and specialized activity within the high-value chemical manufacturing sector when compared to batch processes. Next generation technologies and data are underutilized within these sectors, therefore resulting in unnecessarily long process development times and suboptimal production processes which delivers lower quality product. In this talk, we describe the development of an in-house built optimization and modeling software, which enables researchers to perform sequential experimental design (Fig 1a). The platform facilitates the automation of the flow experiments, collection of data in real-time, and is able to autonomously perform the model fitting in an iterative fashion, with no manual intervention necessary (Fig. 1b).

We highlight the application of the software to optimize a four-dimensional nucleophilic aromatic substitution and a six-dimensional photochemical benzylic bromination. The modular flow platform is coupled with ^1H benchtop NMR for real-time analysis. This talk will demonstrate that the software is easy to implement and flexible to the development constraints, and will thus facilitate early use of models for the identification of the sensitivity of product quality to parameter changes, significantly reducing development time and risks.

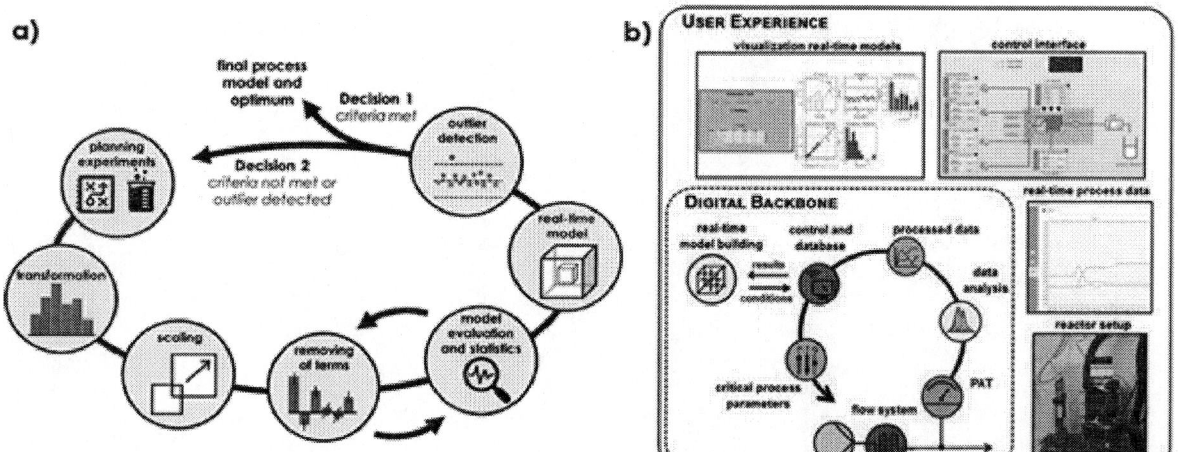

515b Machine Learning-Assisted Autonomous Reactor Characterization: Mass Transfer Coefficients for Oxidative Biocatalytic Reactions

Ajit Vikram, Keith A. Mattern, Shane T. Grosser

Determination of mass transfer coefficients (kLa) is a critical requirement in drug development processes that require aerobic oxidations or other forms of reactions that are limited by supply of gaseous species to a liquid/slurry phase. kLa values depend on several parameters other than reaction conditions, such as reactor geometry, the number of baffles in a reactor, the position, size, and geometry of the agitator. As a result, design of an accurate empirical model for estimating kLa across scales remains very challenging. In addition, the overall kinetics of biocatalytic oxidation reactions are very sensitive to kLa values and thus a slight change in the kLa value across different scales can significantly increase the complexity of the process scale up due to lack of sufficient process information. Therefore, accurate determination of kLa across different scales is often a requirement for development of several drug synthesis processes.

The first part of this talk will focus on developing an accurate predictive model for determination of kLa values across scales ranging from 100 mL to 100 L scale. Specifically, the application of machine learning (ML) based algorithms on archived mass transfer characterization data for developing an accurate kLa prediction model, along with estimation of uncertainty in the ML model predictions will be discussed. The second part of the talk will focus on application of an active learning approach that enables a fully autonomous reactor characterization platform by leveraging ML-guided design of experiments for efficient exploration of the design space by executing minimal number of experiments.

515c Optimization-Based Approaches for Explainable, Automated Chemometric Models

Chrysoula Kappatou, James Odgers, Salvador Salvador Garcia, Ruth Misener

Chemometric models are widely used in the pharmaceutical industry in process analytical technology (PAT) applications. Most of these models are typically built solely on an accuracy objective [1-3], thus not explicitly accounting for model robustness. We refer to robustness in the context of known variability in the training data, e.g., [4-6], rather than robustness with respect to outlier detection [7-8] or downtime in a continuous manufacturing setting [9]. This work proposes an optimization-based approach that incorporates both accuracy and robustness metrics into an optimization problem. We consider accuracy as the goodness of the prediction and propose a moment-matching method that is commonly used in statistics to account for model robustness. Our method includes decisions variables for both the pre-processing and the regression model-building steps. Single and multi-objective formulations, as well as decomposition strategies for the optimization problem are examined. We apply and evaluate our method for industrially relevant case studies from spectroscopy using ENTMOOT [10]. The results illustrate the potential of our approach in automating the decision-making process for building and updating chemometric models, and hence in significantly reducing cost of goods for PAT applications.

References:

[1] Dyrby, M., Engelsen, S. B., Nørgaard, L., Bruhn, M., andamp; Lundsberg-Nielsen, L. (2002). Chemometric quantitation of the active substance (containing C≡ N) in a pharmaceutical tablet using near-infrared (NIR) transmittance and NIR FT-Raman spectra.*Applied Spectroscopy*, *56*(5), 579-585.

[2] Zhang, L., andamp; García-Muñoz, S. (2009). A comparison of different methods to estimate prediction uncertainty using Partial Least Squares (PLS): a practitioner's perspective.*Chemometrics and Intelligent Laboratory Systems*, *97*(2), 152-158.

[3] Bocklitz, T., Walter, A., Hartmann, K., Rösch, P., andamp; Popp, J. (2011). How to pre-process Raman spectra for reliable and stable models?*Analytica Chimica Acta*, *704*(1-2), 47-56.

[4] Igne, B., Shi, Z., Drennen III, J. K., andamp; Anderson, C. A. (2014). Effects and detection of raw material variability on the performance of near-infrared calibration models for pharmaceutical products.*Journal of Pharmaceutical Sciences*, *103*(2), 545-556.

[5] Hetrick, E. M., Shi, Z., Barnes, L. E., Garrett, A. W., Rupard, R. G., Kramer, T. T., Cooper, T. M., Myers, D. P., andamp; Castle, B. C. (2017). Development of near infrared spectroscopy-based process monitoring methodology for pharmaceutical continuous manufacturing using an offline calibration approach.*Analytical Chemistry*, *89*(17), 9175-9183.

[6] García-Muñoz, S., andamp; Torres, E. H. (2020). Supervised extended iterative optimization technology for estimation of powder compositions in pharmaceutical applications: method and lifecycle management.*Industrial andamp; Engineering Chemistry Research*, *59*(21), 10072-10081.

[7] Hubert, M., andamp; Branden, K. V. (2003). Robust methods for partial least squares regression. *Journal of Chemometrics: A Journal of the Chemometrics Society*, *17*(10), 537-549.

[8] Hubert, M., Rousseeuw, P. J., andamp; Vanden Branden, K. (2005). ROBPCA: a new approach to

robust principal component analysis. *Technometrics*, *47*(1), 64-79.

[9] Colón, Y. M., Vargas, J., Sánchez, E., Navarro, G., andamp; Romañach, R. J. (2017). Assessment of robustness for a near-infrared concentration model for real-time release testing in a continuous manufacturing process.*Journal of Pharmaceutical Innovation*, *12*(1), 14-25.

[10] Thebelt, A., Kronqvist, J., Mistry, M., Lee, R. M., Sudermann-Merx, N., andamp; Misener, R. (2021). ENTMOOT: a framework for optimization over ensemble tree models.*Computers andamp; Chemical Engineering*, *151*, 107343.

515d Automated Process in Continuous Pharmaceutical Manufacturing through Real-Time Process Monitoring and Disturbance Detection Using PAT and Online First-Principle Models

Yuma Miyai, Nathan Puryear, Roudabeh Sadat Moazeni Pourasil, Cameron Armstrong, Taylor Raine, Rachel Vallejo, Sherif Abdelwahed, Thomas Roper

With the increased interest in continuous manufacturing (CM) of pharmaceutical processes, implementation of process automation through real-time process monitoring technology plays a critical role in ensuring product quality. The strategy for monitoring and automating continuous synthesis processes through a combined application of process analytical technology (PAT) and online first-principle modelling is discussed for enhanced process robustness. Product concentration predictions were achieved in real-time through PAT implementation using IR and Raman spectroscopy coupled with online chemometric analysis. The process data such as reactor temperatures, starting material concentrations, and stream flow rates were utilized for the computational fluid dynamics (CFD) analysis for calculating the product concentrations based on reaction kinetics as well as mass and heat transfer principles. Process disturbance experiments were conducted for a multi-step flow synthesis process and the changes in the product concentrations were monitored by PAT and the first-principle model. This *in-silico* modelling approach using both the empirical and theoretical models provides robust process monitoring technology which can be utilized in the process decision making such as material diversion and feedback control actions.

515e ML-Ops Platforms and Cloud Computing in Pharmaceutical Development

Jose Tabora

The advantage of model driven decisions in research organizations has long been recognized and applied in the pharmaceutical research and development. From the conceptualization of the Quality by Design framework as a series of mathematical constructs to the application of digital twins, Machine Learning and mechanistic methodologies are widely applicable and efficient tools to construct mathematical models of relevant pharmaceutical processes enabling process optimization, robustness characterization, and failure rate estimation. However, the widespread adoption of these approaches requires a combination of streamlined availability, and institutional know-how to generate, interpret, and incorporate the results in the decision-making process.

This talk addresses different approaches to leverage emerging and stablished tools in Machine Learning and mechanistic software to pharmaceutical development applications from the perspective of options of specific algorithms and computational platforms. Specifically, the integration of cloud-based machine learning operation (ML-Ops) platforms (DOMINO data labs) with elastic cloud High Performance Computing (HPC) providers (Rescale) enables the democratization of domain specific computational tools which would normally require subject matter expertise. In this presentation we demonstrate the deployment of computational fluid mechanics and quantum mechanics workflows that are presented in a simple graphical user interface enabling rapid adoption and integration into the pharmaceutical development process.

515g Digital Assets, Digital Threads, and Digital Twins for Rapid Commercialization of Medicines: A Reference Architecture and Implementation to Enable Pharma 4.0

Pablo A. Rolandi, Matt Potter-Racine

Today's biopharma digital infrastructure is obsolete and fragmented. ELNs are expensive vaults with poor data engineering. PLMs are non-existent or they lack the breadth and depth expected of a true lifecycle management application. While Enterprise Data Lakes have had some success, they can be fragile and are not fully democratized to reach citizen data scientists. Project Portfolio Management may barely exist beyond glorified MS Excel/Project files. Data Science's maximum expression is Random Forests. Machine Learning (e.g., image analysis and NLP) has not reached mainstream use, and In Silico Modeling results may not exist beyond an obscure and fragile shared folder. High Throughput Experimentation's data is untapped due to the drag imposed by data wrangling. Ontologies are missing, rendering end-to-end analyses a mammoth undertaking. Reports, dashboards, may be more commonplace, but GxP validation can be a major burden. With this current landscape, the Pharma 4.0 revolution seems a very daunting task even for the most daring digital innovators in the industry. Finally, cultural norms and organizational constraints can make the transformation even more challenging.

In this work we present a future-view rooted in Amgen's Process Development vast experience in this digital transformation space, and we share with the industrial community, regulatory agencies, and software vendors the many lessons learned. We envision a future of FAIR data assets that are curated and stewarded via effective data governance practices and processes. We imagine a digital thread that fetches these data assets from their sources, and changes them transactionally and programmatically with full traceability and scalability. We recognize that digital twins born out of machine learning and first principles modeling techniques will create cycle times near their theoretical minimum thanks to ubiquitous computation and augmented intelligence. We hypothesize a near-future where these connected assets, threads, and twins will revolutionize CMC authoring, quality processes, and enable major manufacturing and supply efficiencies. Finally, we identify a set of skills, methods, and practices to deliver this vision.

33a Synchronized Mixing, Bubble Size Distribution, and $K_L a$ High-Fidelity Simulation for Optimization, and Scale-up for Benchtop Biofilm Bioreactor

Jorge Lopez, Nima Yazdanpanah, Ali Demirci, Aydin Berenjian

Biofilm bioreactor design, optimization, and scale-up require comprehensive knowledge on all the scale- dependent phenomena, such as mixing performance, oxygen/carbon dioxide mass transfer to/from media, cell biology, impact of shear on cells' performance, and multiphase multiscale systems. Biofilms represent a natural form of cell immobilization, and an increased population of microorganisms in the bioreactor for the production of value-added products. The designed bioreactor for this project includes a set of Plastic Composite Supports (PCS) on the agitator shaft, which certainly influences the agitation performances and mixing profile.

The overall mixing performance, shear rate impact on cells' behavior, distribution, and population balance of gas bubbles, are characteristic indication of the system at steady-state. However, the mass transfer, $k_L a$, oxygen consumption and carbon dioxide release, cell growth, and biofilm characteristics are transient and scale dependent. On the equipment consideration side, the heat transfer area, sparger design and aeration rate (i.e., bubble size), volume to height ratio, radial/axial dispersion of fluid flow and energy distribution are other nonlinearities in the systems. Studying these aspects individually cannot provide a holistic view on the bioreactor performance, and any tools and knowledge developed this way is incomplete and cannot be used for robust process development, optimization, and efficient scale-up.

In this study, we developed and adopted a synchronized mixing, bubble size distribution, and $k_L a$ high-fidelity simulation (computational fluid dynamic-CFD) for a customized biofilm bioreactor. All physics and system nonlinearities were considered at different scales. Transient two-phase flow simulations were conducted were a S-Gamma population balance was included to consider the breakup and coalescence of gas bubbles.

The model also includes the interaction of the gas bubbles with the mixing element of the bioreactor, and its effect on the mass transfer and $k_L a$. Furthermore, a unique impeller design was customized and optimized for a biofilm bioreactor. Different design characteristics of the impeller (e.g., length, number of elements, etc.), and bioreactor operating conditions, such as RPM, were evaluated in the performance of the mixing properties (i.e., pumping and mixing number) and mass transfer characteristics (i.e., $k_L a$, distribution of the oxygen/carbon dioxide in the biorector, and gas bubble size distribution).

33b Process Development and Scale-up of an Enzyme Removal Strategy for Drug Substance Manufacturing

Victoria Zhang, Jonathan P. McMullen, Jungchul Kim, Kevin Sirk, Alexandra Sun, Daniel DiRocco, Joshua N. Kolev

Biocatalytic transformations have become increasingly prevalent in the pharmaceutical industry enabling the highly selective and sustainable synthesis of active pharmaceutical ingredient (API) intermediates. However, the removal of these enzymes in downstream processing to ensure the isolation of high quality intermediates not only poses a challenge during development at benchtop scale but also can often be a bottleneck in the long term manufacturing process. Therefore, to ensure successful tech transfer to manufacturing, an in-depth understanding of process sensitivities and appropriate scale-up parameters is critical when evaluating the various methods of enzyme removal during process development.

This presentation will discuss the various enzyme removal strategies that were explored during the development and optimization of an enzymatic hydroxylation process to isolate a drug substance intermediate. The intermediate is typically recovered by extraction with an organic solvent from the aqueous system; however, the presence of denatured enzymes can often obscure the liquid-liquid interface. The enzymes can alternatively be filtered off with the addition of a filter aid but enzyme filtration can often be inefficient and difficult to predict upon scale-up without proper development. Benchtop experiments were performed to explore liquid-liquid extraction and depth filter cartridges as well as the filtration (with the addition of filter aid) of not only the enzyme itself but also both the enzyme and drug substance intermediate. After assessing a variety of metrics including process efficiency and isolated intermediate quality, filtration of the enzyme with the addition of a filter aid was found to be optimal for our system. This process was optimized at benchtop scale before it was transferred initially to the kilo-scale lab and then to the pilot plant. Additionally, a process fingerprint was collected with a fiber optic scale in the pilot plant to demonstrate the robustness of the enzyme removal approach developed.

33c Scale-Dependency Assessment Strategy to Translate Proven Acceptable Ranges from Small to Commercial Scale

Mary am Ende, Jacob Santos-Marques, David Am Ende, Matthew Burk, Fern Sinclair, Jerry S. Salan, David Leahy, Lopa Bakrania

Regulators have raised concerns about the applicability of proven acceptable ranges (PARs) developed on small scale to commercial manufacturing processes. These concerns are forged out of potentially detrimental impact of scale on the desired attributes of the drug substance or product. Requests have been made to verify PARs with data generated at commercial scale. It is not always feasible to manufacture large scale batches, therefore alternative strategies are needed.

In this presentation, the criteria for judging scale-(in)dependence of synthetic processes will be discussed. A decision tree will be shared for guiding the scale-dependence evaluation of each process parameter to define whether the PARs can be directly translated from small to large scale or if further work is needed. Finally, specific examples of scale-independent and scale-dependent parameters will be discussed with strategies (e.g. stressing studies, computational modeling) to alter the final categorization. These approaches address regulatory concerns around the applicability of the established PARs on commercial scale and process without compromising the quality standards for the pharmaceutical drug substance and drug product.

33d Islatravir Crude Isolation Development and Scale-up: Late-Stage Shift from Filter Dryer to Centrifuge to Mitigate Risks from Enzyme Filtration

Brittany Dobson, Maggie Miller, Daniel Bishara, Matthew Gunsch, Michelle Cleary, David Bell, Aisling Prendergast, Ivan Lee, Darryl Chang, Sandra A. Robaire, Nicholas Rogus, Theodore Furman, Gregory Hughes, Elizabeth Fisher, Kevin Maloney, Junyong Jo, Paula Archbold

Islatravir is a nucleoside reverse transcriptase inhibitor small molecule targeted for HIV. Islatravir is synthesized through a biocatalytic cascade using nine enzymes and has a single isolation. As such, the filtration was known to be challenging from the start of process development due to the presence of enzymes and proteins. Denatured enzymes can settle and clog the filter, resulting in elongated filtration times and limited rejection of residual proteins. Reducing enzyme loading levels and adding a filter aid were explored during early development to improve filtration efficiency in the filter dryer. These process modifications resulted in initial success as the process was scaled up from bench scale to the pilot plant. During further scaleup, the filtration progress became significantly slower, requiring days to complete the filtration. This highlighted both operational and quality risks long term for the process. This led to the decision to pursue a late-stage process change from a filter dryer to a centrifuge. This prompted the team to initiate rapid lab development to understand and mitigate concerns around differences between the two pieces of equipment pertaining to compressibility, washing, and temperature control. Centrifugation can be challenging to adequately evaluate at the lab scale due to the dependence on basket size and changing liquid height as the driving force. Dynochem modeling was leveraged to bridge the gap between bench and pilot scale to inform operating conditions for initial pilot scale batches. The pilot scale demonstration investigated a range of operating parameters and derisked the initial processing concerns, clearing the way for implementation of the process at commercial scale. The work culminated in the successful utilization of the centrifuge at manufacturing scale, achieving reproducible filtration performance across multiple batches, isolated product of desired quality, and additionally a significant reduction in drying cycle time. This talk will focus on the decision to change equipment, the risk assessment and considerations, as well as the modeling and implementation of the centrifuge across scales to enable a successful enzymatic filtration.

566c *N-Glycanyzer*: An Integrated on-Line Process Analytical Toolkit for Enabling Continuous Biologic Manufacturing

Aron Gyorgypal, Shishir Chundawat

The biopharmaceutical industry is transitioning towards the adoption of continuous biomanufacturing practices that are often more flexible and efficient than traditional batch processes. Federal regulatory agencies like the FDA are further urging use of advanced Process Analytical Technology (PAT) to analyze the design space to increase process knowledge and enable high quality biologics production. Post-translational modification of proteins, such as N-linked glycosylation, are often critical quality attributes that affect biologics safety and efficacy, requiring close monitoring during manufacturing. Here we have developed the *N-Glycanyzer* as an integrated PAT tool that combines on-line sampling and sample preparation with a mobile HPLC unit allowing for automated and near-real time measurement of monoclonal antibody (mAb) glycosylation (1). This toolkit can be used to understand the temporal changes of glycosylation during fed-batch and perfusion based mAb manufacturing allowing for tighter control of the glycan profile, leading to less variability during upstream bioprocessing. We will highlight the key unit operations and optimization techniques used to build the *N-GLYcanyzer* PAT tool and provide an example of its use for monitoring a fed-batch upstream process using mammalian cells for producing a commercially relevant Trastuzumab biosimilar.

1. Gyorgypal, A. andamp; Chundawat, S. P. S., 'Integrated Process Analytical Platform for Automated Monitoring of Monoclonal Antibody N-Linked Glycosylation,' *Analytical Chemistry.*analchem.1c05396 (2022). DOI: 10.1021/acs.analchem.1c05396. Weblink: https://pubs.acs.org/doi/10.1021/acs.analchem.1c05396

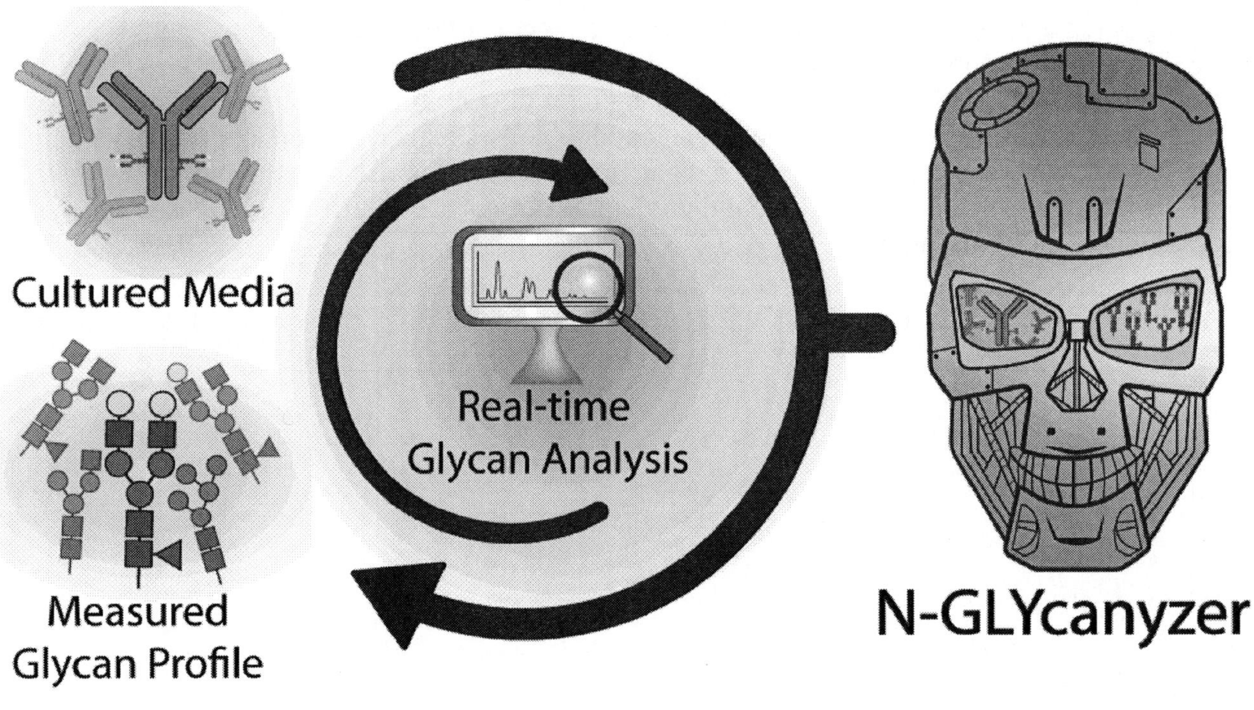

33f Antibody Production Against Camptothecin- Derived Small Molecules :a Tool for Developing Pharmacokinetic Studies and Dose Management Chemotherapy

Tahereh Zarnoosheh farahani

Introduction
Camptothecin compounds (such as irinotecan or topotecan) are one of the commonly used and most operative chemotherapeutic drugs for cancer therapy. The active form of all camptothecin derivatives is
SN38 released by liver esterase enzyme as the super-cytotoxic form of camptothecin families. Furthermore, in recent years, SN38 itself is also used as a payload in new generation of targeted drug delivery systems termed ADCs technology. Pharmacokinetic studies (PK) and side effect management of
the aforesaid cancer medicines have been remained a challenge for optimization of patient safety and improving medical practice. To overcome this challenge, developing simple analytical method, non-invasive and amenable to routine clinical laboratory is the main goal of our present study.
Method
For this purpose, The SN38–20-O-glycine and SN38–20-O-glycine-NHS was synthesized and confirmed
by mass, 1 H -NMR, and 13 C -NMR spectroscopic techniques. Then two different immunogens were prepared and characterized, including SN38 conjugated with keyhole limpet hemocyanin (KLH) protein
based on linking of SN38 either via its amine-containing SN38–20-O-glycine and SN38–20-O-glycine-
NHS derivatives with the tyrosine amino acid residues of KLH by classical glutaraldehyde coupling reaction. The pertinence and efficacy of the coupling reactions of haptens (SN38-KLH and SN38NHS-
KLH) were confirmed and characterized by ultraviolet (UV) spectrophotometry and then used for the immunization of animals. Antibodies production and isotypes determination were done by enzyme-linked
immunosorbent assay (ELISA) and then purified either using handmade SN38 affinity or HiTrapby Protein G prepacked columns.
Results
Immunization of animals with the SN38NHS-KLH generated extraordinarily high anti-SN38 titers that
were up to-fold higher than those immunized by SN38-KLH. The animal's antiserum that showed the highest affinity was selected. The collected antiserum (polyclonal and monoclonal antibodies) had very
high affinity to its cognate ligand with high affinity and high specificity among other irrelevant drug controls.
Conclusion
The produced polyclonal and monoclonal antibodies are valuable for the development of highly sensitive
and selective immunoassays for camptothecin- containing cancer medicines. The sensitivity and

specificity of the ELISA should provide a useful tool for developing pharmacokinetic studies, dose monitoring and side effect management of camptothecin-derived small molecules.
Key words

SN38, cancer chemotherapy, pharmacokinetic studies, anti-SN38 antibodies, ELISA, immunoassay,

Emerging Microfluidic Platforms for Crystallization Process Design: From Academic Research to Commercialization through ETC

Meenesh Singh

Crystalline materials are used extensively in a variety of healthcare- and energy-relevant technologies, e.g., controlled-release tablets, highly efficient energy storage materials, low-temperature catalysts, low-cost separation for carbon capture and water treatments, and materials of extreme environments. In the case of biopharmaceuticals, it currently takes around 10 years and billions of dollars to bring a new drug to market after an API molecule is invented to treat a disease. A significant portion of that decade is invested in
process development, where companies screen different polymorphic forms of APIs, and develop robust processes to manufacture the stable form with the acceptable physical properties to turn that into tablets. Commercial microtiter plates upto 1584-wells and holding volumes as low as 1 μL are frequently used to evaluate ~192 conditions per run to screen salts, co-crystals, polymorphs, and solubility. These devices currently in market run into problem due to depletion of supersaturation and that is the major reason for
setbacks during the drug product development. To combat this issue, we have developed a high-throughput continuous-flow microfluidic device that ensures well-mixed and constant supersaturation conditions by continuously supplying fresh API solution. Such supersaturation-controlled screening device has never been developed before because most of the microfluidic devices lack the mechanism to trap API crystals, which is needed for analysis and measurement of polymorphism, nucleation, and growth.

Our pre-competitive project with ETC resulted in an advanced microfluidic device, which is currently at TRL 7. The ongoing project with ETC and eN-RAMPS will help in the automation of the device, which will bring our technology to TRL 8+ level (closer to commercialization). The objective is to develop and deploy a standalone fully-automated microfluidic system to measure crystal growth rate and nucleation rate, screen polymorph, and assess solid-liquid and liquid-liquid equilibrium, which will be seamlessly coupled
with a graphical user interface to characterize and optimize the crystallization process quickly and reliably. The fully-automated microfluidic device will also help biopharmaceutical companies comply with an FDA initiative to move from batch-based production to continuous manufacturing, which will spur advancements
in drug production and lower the cost.

The Compact HPLC Project: A Collaboration between Axcend and the Enabling Technologies Consortium

Ray West

This presentation will review the collaboration between Axcend and the Enabling Technologies Consortium (ETC) to develop a compact HPLC. The effectiveness of the collaboration between eight ETC members and Axcend to define, develop, supply, and evaluate a viable compact HPLC system will be highlighted. The opportunity to develop technology and provide a solution with broad representation across the pharmaceutical industry has been invaluable. This highly productive project, which followed an earlier ETC project that addressed on-line sampling for HPLC analysis, has enabled Axcend's technology platform to be modified and further developed to meet the specific needs for on-line characterization of chemical processes in a space-limited chemical fume hood. Beta-testing of the compact HPLC system by ETC members revealed additional applications in a variety of other areas of interest in pharmaceutical discovery, development, processing, and quality control.

Next Generation Models for Crystal Morphology Prediction – a Collaboration between ETC and Ucsb

Michael F. Doherty

Crystals almost always grow as faceted particles exposing a small number of chemically distinct faces on their surface. Since each face exposes different chemical groups, different faces interact differently with the solution environment, which is why crystals of the same solute with different morphologies behave differently (e.g., dissolve differently). Crystal shape is also important. Generally, needles and plates are problematic for processing and tableting. Therefore, crystal morphology and shape are important material characteristics and there is significant value for models that are capable of simultaneous prediction of crystal shape and morphology. Such models may be used to perform computer experiments that help guide experiments to identify preferred solvents, temperatures, supersaturations and other process conditions that lead to desired crystal forms. To be successful in a process development workflow, morphology prediction models must be fast (compute times measured in minutes not days or weeks) and yet remain faithful to the fundamental solid-state chemistry and surface physics governing the growth processes during evolution of particle morphology.

In recent years, a collaboration between our team at UCSB and the ETC Consortium has helped advance the model development to predict morphology and shape of organic crystals grown from solution. The models are based on surface integration kinetics as the rate determining step. The key features that have been or are continuing to be addressed include: (1) importance of optimization of the hydrogen positions in the crystal structure file, (2) use of the COSMO-SAC property model to estimate solvent-modified bond energies between growth units on the step edges that flow across crystal faces, (3) models for calculating the density of kink sites along step edges at high supersaturation (i.e., liberation from equilibrium Boltzmann statistics), (4) nonlinear step velocity models suitable for use at high supersaturation, (5) importance of validating the classical atom-atom force field from among those available (e.g., GAFF, CLP, Lifson, etc.). These advances either have been or are currently being incorporated into our digital design aid ADDICT. Sponsor testing and feedback has been a valuable source of inspiration.

User-Friendly Population Balance Model with Data Integration for the Digital Design of Crystallization Systems

Zoltan Nagy

Industry-wide precompetitive collaborations have been increasingly promoted by the pharmaceutical industry [1]. The mission of the Enabling Technologies Consortium (ETC) is to enable rapid, pre-competitive collaborative development of new enabling technologies for the pharmaceutical industries. Precompetitive collaborations have advantages beyond the financial benefits (of leveraged risk) to their participants, as they can also be used to deliver products with tailored features valuable to their users in contrast to the development of tools with broad features generally applicable to all industries but not necessarily ideal for any of them. For many ETC projects, future users are directly involved in product development, at times serving as high level technical advisors and beta-testers. In this work present the product of one of the first ETC projects, which has resulted in a new population balance model (PBM) based interactive, user-friendly crystallization software for pharmaceutical research and development. The key scope of the Crystallization Simulation and Visualisation (CrySiV) software tool is to perform process modeling and simulation of industrial crystallization processes by regressing model parameters from the experimental data [2]. Consequently, through in-silico experiments the software can enable better understanding of key correlations between critical quality attributes (CQAs) and critical process parameters (CPPs) and ultimately promote the efficient and rapid digital design of crystallization systems. CrySiV is aimed to merge the state-of-the-art in crystallization modeling, simulation, parameter estimation and optimization into a user-friendly interface. It allows efficient input of typical experimental data, guides the user to perform efficient parameter estimation and enables in silico investigations and digital design of crystallization and integrated crystallization and wet-milling systems, for both batch and continuous processes. The talk describes the key features of the CrySiV software, introduces the 1D and 2D PBM framework used in the model and provide examples of how the tool can facilitate rapid crystallization design in the case of a batch combined cooling, antisolvent crystallization, investigate transition of batch to continuous crystallization processes and perform the digital design of an integrated crystallization wet milling system. The successful examples demonstrate the benefits of in silico design of experiments and digital design of crystallization systems, through saving experimental time and material, and speeding up industrial crystallization development, thereby CrySiV promotes the broader adoption of digital technologies in modern pharmaceutical manufacturing.

[1] J. Welch, C.; M. Hawkins, J.; Tom, J., Precompetitive Collaboration on Enabling Technologies for the Pharmaceutical Industry, Org. Process Res. andamp; Dev., 18 (4), 481–487, 2014.

[2] B. Szilagyi, W.-L. Wu, A. Eren, J. Mackey, S. Kshirsagar, E. Szilagyi, I. Ostergaard, K. Sinha, L. Mlinar, D. Pohlman, J. Chen, N. Nere, B. Moussa, M. Lovette, S. Black, A. Jawor-Baczynska, H. Li, B.-S. Yang, E. Irdam, D. Patience, R. McKeown, D. Juboor, M. Ketchum, D. Green, C. Polster, C. Burcham, D. Jarmer, J. Miles Merritt, L. Codan, J. Schoell, A. Cote, E. Sirota, Y.C. Liu, K. Girard, S. Kulkarni, Y. Yang, J. L. Quon, Z. K. Nagy, Cross-pharma collaboration for the development of a simulation tool for the model based digital design of pharmaceutical crystallization processes (CrySiV), Cryst. Growth andamp; Des., 21 (11), 6448-6464, 2021.

273b Pharmacy on Demand: A Case of Arrested Development or Flourishing Adolescence

Luke Rogers

Pharmacy on Demand is an initiative to realize mobile production skids that utilize continuous manufacturing unit operations to produce both active pharmaceutical ingredients and final drug products. This is in direct response to rising drug prices and centralized supply chains. These proof-of-concept units placed an emphasis on automation, reduced form factor, and modularity. Throughout the lifetime of the PoD initiative, substantial advances have been made in complex reaction telescoping, real time formulation, reaction engineering, and pumping technologies. Our vision of an end state for the Pharmacy on Demand project, however, has taken on a longer view. We have begun to re-imagine how we both develop syntheses and prototype equipment with respect to regulatory guidelines, and in particular, how we address life-cycle management. We have begun to apply these updated workflows to the new additions to our product portfolio in conjunction with best practices from other industries on supply chain management. We believe PoD is growing up. We believe we have emerged from a proof-of-concept mindset to a flourishing adolescence as we prepare for a suite of commercial regulatory filings in 2022 and beyond. We wish to share the new processes we have developed for these molecules as well as generalized workflows to decrease the transition time from batch discovery to continuous production mode.

References:

1. Mascia, P. L. Heider, H. Zhang, R. Lakerveld, B. Benyahia, P. I. Barton, R. D. Braatz, C. L. Cooney, J. M. B. Evans, T. F. Jamison, K. F. Jensen, A. S. Myerson and B. L. Trout, *Angew. Chem., Int. Ed.*, **52**, 12359–12363
2. P. Cole, J. M. Groh, M. D. Johnson, C. L. Burcham, B. M. Campbell, W. D. Diseroad, M. R. Heller, J. R. Howell, N. J. Kallman, T. M. Koenig, S. A. May, R. D. Miller, D. Mitchell, D. P. Myers, S. S. Myers, J. L. Phillips, C. S. Polster, T. D. White, J. Cashman, D. Hurley, R. Moylan, P. Sheehan, R. D. Spencer, K. Desmond, P. Desmond and O. Gowran, *Science*, 2017, **356**, 1144–1150
3. Adamo, R. L. Beingessner, M. Behnam, J. Chen, T. F. Jamison, K. F. Jensen, J.-C. M. Monbaliu, A. S. Myerson, E. M. Revalor, D. R. Snead, T. Stelzer, N. Weeranoppanant, S. Y. Wong and P. Zhang, *Science*, 2016, **352**, 61–67
4. Rogers, L., Briggs, N., Achermann, R., Adamo, A., Azad, M., Brancazio, D. et al., Continuous Production of Five Active Pharmaceutical Ingredients in Flexible Plug-and-Play Modules: A Demonstration Campaign. Organic Process Research andamp; Development 2020, 24 (10), 2183-2196.
5. Imbrogno, J., Rogers, L., Thomas, D., Jensen, K.F. Continuous Purification of Active Pharmaceutical Ingredients Utilizing Polymer Membrane Surface Wettability. Commun., 2018, 54, 70-73.
6. Srai JS, Graham G, Hennelly P, Phillips W, Kapletia D, Lorentz H. Distributed Manufacturing: A new form of localised production?, International Journal of Operations and Production Management. 2020 40 (6), 697-727
7. Harrington, T.S., Phillips, M.A., Srai, J.S. 2016, *Reconfiguring global pharmaceutical value networks through targeted technology interventions*, International Journal of Production Research, http://dx.doi.org/10.1080/00207543.2016.1221541
8. https://www.fda.gov/regulatory-information/search-fda-guidance-documents/q13-continuous-manufacturing-drug-substances-and-drug-products

272a Facility Design for Continuous Clinical Supply Facilities and Laboratories

Christopher Vitale, Jie Chen, Moiz Diwan, James Ratway

Abstract: Efficient and agile process development and manufacturing for clinical supplies requires that certain lab and scale-up facility infrastructure and process flexibility concepts be utilized. The facility layout and design features provide the tools needed by the development team and can play an important role in the success of a process development campaign. In this work, we will review details of two recent facility upgrades at AbbVie's API Pilot Plant and Process Randamp;D Laboratories that help to achieve efficient continuous processing. Case studies of continuous processing setups will be presented to demonstrate how the facility infrastructure lends itself to a more optimized and agile process development campaign.

Disclosures:

All authors are employees of AbbVie and may own AbbVie stock. The design, study conduct, and financial support for this research were provided by AbbVie. AbbVie participated in the interpretation of data, review, and approval of the publication.

273c A Hybrid Liquid Phase Peptide Synthesis (LPPS) Facility

Darragh McDonagh

A cGMP manufacturing facility employing both small volume continuous (SVC) and traditional batch equipment to produce synthetic peptides. Feed make-up vessels, SVC skids operating in fume hoods, nanofiltration purification skids, surge vessels, precipitation vessels and agitated filter/dryers are all employed to enable production at metric ton scale. The distributed control system controls and monitors the process in conjunction with on-line process analytics and electronic batch records. Finally, a digital batch genealogy model is used to streamline batch release.

273d Digitalization in Pharmaceutical Development

Jaquan Levons

Pharmaceutical development is a rich and unique opportunity for value through digitalization. The business processes in our industry are not typically optimized for the rapid and agile approach to innovation that aggressive digitalization requires. Successful digital projects have certain attributes that align them with business processes, but lag in terms of progress towards modernization. Key aspects of development and of the business processes end up manifesting challenges to meaningful change. In this presentation we will review key successes in our organization, key challenges, and potential solutions to executing a contemporary digital strategy in product development.

273a Modular Flexible Manufacturing - Enabling and Accelerating Specialty Product Development and Commercialization

Athanas Koynov, Celeste Frankenfeld

The "traditional" pharmaceutical product development and commercialization paradigm involves formulation development (and early phase clinical supply manufacture) carried out at a development site, followed by commercialization, and launch, at a separate commercial site. In this scenario both sites feature equipment associated with common drug product manufacturing unit operations (e.g., powder conveying, blending, roller compaction, tableting, etc., in the case of oral solid dose formulations). Commercialization typically consists of the tech transfer and scale-up from the development to the commercial equipment train.

This approach fits the pipelines of the past, which were dominated by large-volume, small-molecule, orally delivered drug products. A future state characterized by small-volume, targeted therapies, based on a diversity of modalities, formulation types and delivery routes, however, will likely necessitate the redesign of the development and commercialization paradigm as well.

This talk will present the concept of a network of flexible and versatile sites capable of rapidly tech transferring and commercializing any set of unit operations through the use of self-contained, mobile, autonomous manufacturing units (PODs). The implementation of the concept will be demonstrated using the example of an implantable, long-lasting parenteral product – a formulation requiring non-typical equipment and manufacturing space requirements – without an installed commercial manufacturing base, and without a following pipeline.

A Digital Paradigm for mRNA Vaccine Development

Aparajita Dasgupta

Despite the advances and increased usage of process analytical technology (PAT) in biomanufacturing pipelines, current industry practices for bioprocess development and characterization have yet to suitably integrate the vast amounts of data that is generated from these tools. With new modalities such as mRNA vaccines becoming crucial tools that have proven their potential to target global health issues rapidly and effectively, it is imperative that we develop our bioprocessing platforms to allow for fast development, prototyping, and deployment of new products. Integrating a model-based strategy within current workflows may reduce the number of experiments needed to characterize these processes, thus reducing timelines while simultaneously improving product quality. In this talk, we discuss our efforts on this front, primarily focusing on our efforts within the space of mRNA therapeutics to integrate digital twins within our bioprocess development pipelines within Pfizer. We discuss our multimodal efforts using machine learning and mechanistic modeling to move towards a Pharma 4.0 paradigm and reflect on current and future industry practices that we believe may enable a rapid transformation within this space.

700a Two-Dimensional Population Balance Model for Twin-Screw Wet Granulation: Model Development and Calibration Based on Particle Size Distribution and Porosity

Kensaku Matsunami, Ana Alejandra Barrera Jimenez, Michiel Peeters, Michael Ghijs, Daan Van Hauwermeiren, Thomas De Beer, Ingmar Nopens

In the pharmaceutical industry, there is a transition ongoing from conventional batch to novel continuous manufacturing. Twin-screw wet granulation (TSWG) is one of the emerging technologies in continuous manufacturing of solid oral dosages. Quite a few researchers have been working on mechanistic modeling of TSWG to predict granule quality attributes as well as to understand granulation mechanisms which occur spatially along the twin-screw granulation barrel. Among multiple types of mechanistic models, population balance models (PBMs) have been widely applied to TSWG for the computation of granule quality attributes, e.g., granule size distributions (GSDs).

One-dimensional (1D) PBM has been the most popular approach for the application to the pharmaceutical industry because of data availability. Existing studies reported that GSDs were successfully simulated by using 1D PBM[1–4]. However, 1D PBM can only predict a single quality attribute, i.e., GSDs, and has difficulties in capturing some granulation phenomena, e.g., consolidation. With 1D PBM, it is impossible to simultaneously assess other quality attributes, e.g., porosity, whereas this is also important for the final product quality, e.g., tablet dissolution. This fact proves that the development of multi-dimensional PBMs is valuable to the industry.

One of the biggest challenges to develop multi-dimensional PBMs is the difficulty of obtaining sufficient and appropriate experimental data. Whereas multi-dimensional PBMs have been proposed by numerous authors[5–9], all of these models have yet to be calibrated for the formulations containing active pharmaceutical ingredients (APIs). Multi-dimensional PBMs mostly consider the items that require advanced and detailed measurements, e.g., liquid distribution.

Therefore, this work proposes a new two-dimensional (2D) PBM for TSWG focusing on GSD and granule porosity. Granulation experiments were performed for three formulations with different liquid-to-solid ratios using the TSWG of the ConsiGma™-25 (GEA Pharma Systems, Collette™, Wommelgem, Belgium), which consists of one wetting zone, two kneading zones, and size control elements. After each zone, granule porosity at five different size classes was characterized, in addition to GSDs. A 2D compartmental PBM was developed based on the experimental data. Pure aggregation was assumed for the wetting zone, where aggregation kernels were defined based on the kernels used in the 1D PBM[1, 2]. In the kneading zones, three phenomena were taken into account for the model: aggregation, breakage, and consolidation. The simulation results using the developed PBM were compared with the experimental data.

From the experimental data, the impact of consolidation was observed in terms of reduced granule porosity after the kneading zones. This impact could be reflected in the developed PBM. In addition, model calibration of the wetting zone was successful for multiple settings of liquid-to-solid ratio. This means that the kernels developed for the 1D PBM can be applied to the 2D PBM because they already capture important phenomena sufficiently.

Overall, this work demonstrated the applicability of the 2D PBM for the pharmaceutical industry. The developed model can provide information of granule porosity as well as GSDs, which can increase the value of TSWG models. In the future, further calibrations with different formulations can make the developed 2D PBM more valid and generic.

Acknowledgment: The authors would like to acknowledge UCB for their financial support and fruitful collaboration in this project.

References:

1. D. Van Hauwermeiren, M. Verstraeten, P. Doshi, M. T. am Ende, N. Turnbull, K. Lee, T. De Beer, I. Nopens, On the modelling of granule size distributions in twin-screw wet granulation: Calibration of a novel compartmental population balance model, Powder Technology 341 (2019) 116–125.

2. A. A. Barrera Jiménez, D. Van Hauwermeiren, M. Peeters, T. De Beer, I. Nopens, Improvement of a 1d population balance model for twin-screw wet granulation by using identifiability analysis, Pharmaceutics 13 (2021) 692.

3. H. Y. Ismail, S. Shirazian, M. Singh, D. Whitaker, A. B. Albadarin, G. M. Walker, Compartmental approach for modelling twin-screw granulation using population balances, International Journal of Pharmaceutics 576 (2020) 118737.

4. H. Liu, S. C. Galbraith, S. Y. Park, B. Cha, Z. Huang, R. F. Meyer, M. H. Flamm, T. O'Connor, S. Lee, S. Yoon, Assessment of spatial heterogeneity in continuous twin screw wet granulation process using three-compartmental population balance model, Pharmaceutical Development and Technology 24 (1) (2019) 105–117.

5. K. F. Lee, S. Mosbach, M. Kraft, W. Wagner, A multi-compartment population balance model for high shear granulation, Computers \andamp; Chemical Engineering 75 (2015) 1–13.

6. H. Y. Ismail, M. Singh, A. B. Albadarin, G. M. Walker, Complete two dimensional population balance modelling of wet granulation in twin screw, International Journal of Pharmaceutics 591 (2020) 120018.

7. C. Sampat, Y. Baranwal, R. Ramachandran, Accelerating multidimensional population balance model simulations via a highly scalable framework using GPUs., Computers \andamp; Chemical Engineering (2020) 106935.

8. L. G. Wang, S. U. Pradhan, C. Wassgren, D. Barrasso, D. Slade, J. D. Litster, A breakage kernel for use in population balance modelling of twin screw granulation, Powder Technology 363 (2020) 525–540.

9. L. G. Wang, J. P. Morrissey, D. Barrasso, D. Slade, S. Clifford, G. Reynolds, J. Y. Ooi, J. D. Litster, Model driven design for twin screw granulation using mechanistic-based population balance model, International Journal of Pharmaceutics 607 (2021) 120939

700b Machine Learning Approaches for the Prediction of Powder Behavior of Pharmaceutical Formulations from Physical Properties of the Active Pharmaceutical Ingredient

Laura Pereira Diaz, Stephanie Marchal, Patrick M. Piccione, Cameron Brown, Alastair J. Florence

Background

Formulation and process development activities are necessary to make a drug substance into a drug product – yet can be a time-consuming, iterative process. Artificial intelligence (AI) and machine learning (ML) have emerged as potential tools to accelerate the transition from formulation development to manufacturing and thus, the use of digital design and data-driven models provides the prospect to accelerate these important development steps (1). Changes in formulation from variation in excipients or their composition can impact bulk properties such as powder flowability and therefore risk impacting subsequent manufacturing processes (2). More subtle changes, for instance in the physical properties of the active pharmaceutical ingredient (API), such as particle size and shape, can also influence the manufacturability of the drug product (3, 4).

In the present work, we assess ML models to enable rapid decision-making on the viability of pharmaceutical formulations for continuous direct compression (CDC), hence minimizing the time and experimental resources needed to make this decision. 35 excipients and 17 APIs, combined in 75 pharmaceutical formulations, were analyzed and included in the training dataset of the ML classification models. The models classify the formulations into viable and non-viable categories depending on their flow properties, assessed with respect to the suitability for CDC. This abstract shows the preliminary results of an ongoing study.

Methods

To make reliable predictions using ML models, adequate training data are required to describe the range of variability encountered in the objective. Therefore, the data were first generated experimentally, followed by data curation, model training, and validation.

First, particle size and shape distributions of the single compounds for a series of test formulations were included in the training set as the independent variables. Particle shape (convexity, elongation, length, area, perimeter, sphericity, and aspect ratio) was measured using static image analysis performed by a Morphologi G3 (Malvern), and particle size distribution (D10, D50, D90) was measured using both laser diffraction (MasterSizer 3000, Malvern) and dynamic image analysis (QICPIC, Sympatec). The flow function coefficient (FFc) of the test formulations was included in the training dataset as the dependent variable (response). The FFc was measured with the ring shear tester created by Dietmar Schulze at a normal consolidation stress of 1000 Pa (relevant pressure of the CDC manufacturing line under consideration). The formulations were classified as viable or non-viable if their FFc was greater or smaller than 5, respectively, based on equipment needs determined in-house. The data curation process involved filtering and preparing the final dataset that will go into the training step, i.e., Pearson correlation coefficient was calculated to remove the highly correlated features.

For the model training, initially, Principal Component Analysis (PCA) and Louvain clustering were performed as unsupervised algorithms for multivariate data analysis and to identify smaller groups of variables to investigate the hidden patterns within the data (5, 6). Likewise, Louvain clustering is used to detect groups of data in a large dataset, so the similarities between the formulations can be analyzed (7). PCA was calculated using the python package sci-kit learn decomposition and visualized using the python library matplotlib. Louvain clustering was calculated and visualized using the Orange Data Mining software.

Subsequently, supervised learning algorithms were used to build a model that classifies formulations into viable and non-viable classes (threshold FFc = 5). The first step was the selection of the algorithm, using precision as the metric to compare model effectiveness. Precision is the ratio of the true positives to the total instances predicted as positives. High precision ensures that the formulations that are predicted as viable are indeed viable and therefore, suitable for CDC. The main reason to choose this metric is to avoid false positives (non-viable formulations predicted as viable) which would lead to waste of time and resources and jeopardize customer confidence. The best performing algorithm was selected among the following: random forest, support vector machines, gradient boosting, AdaBoost, k-Nearest Neighbors, logistic regression, neural network, and Naïve Bayes. Due to data availability constraints, the data was sampled using leave-one-out cross-validation to compare the performance of the classification algorithms, where one formulation was used for testing at a time, and the remaining formulations were used for training. This process was repeated for all the formulations included in the dataset. After this stage, the confusion matrix was computed from which the precision was calculated. Finally, the model was evaluated using an external set generated experimentally.

Results

The dataset included 75 formulations of different compositions and different drug loadings. The concentration of the formulation, particle size and shape descriptors of the API, 35 different excipients, and their FFc are included in the dataset. The calculation of the Pearson correlation coefficient helped decide what particle size and shape descriptors should be included in the training dataset. Setting a threshold of 0.9, descriptors that were highly correlated were removed from the analysis, leaving only 17 variables from the initial 35 size and shape descriptors. After this process, the data are curated and prepared for training.

The unsupervised learning algorithms failed since they did not show clusters of data based on the FFc of the formulations and thus, did not provide a useful classification method. Among the classification algorithms, gradient boosting achieved the best performance, exhibiting the highest precision score (0.896). Feature importance analysis was calculated based on the decrease of the precision for gradient boosting, denoting that the most important variables towards the result of the prediction were the FFc of the API and its particle size distribution Morphologi G3 D50 value. Furthermore, an external test was conducted to validate the performance of the model on unseen materials. More data is being generated to complete the dataset.

Implications

The ML classification model's initial 89.6% precision in prediction of viability of a formulation for CDC based simply on particle size and shape measurements, FFc of single components, and composition of the formulation is a promising indication that this technique could have routine application in rapid assessment of manufacturability of novel formulations. This would allow rapid decision making in early-stage development when time and the amount of API are at a premium. The precision of the model is acceptable; however, it can be improved with the availability of more data from wider training set of materials and work is continuing to extend the training set with a wider range of material attributes, composition and formulation performance and improve the general applicability. The model could ultimately be adapted to different FFc thresholds, adaptive to the considered manufacturing process. Its extension to other formulations attributes (e.g., wall frictions) will be further investigated. The model could also be extended to inform formulation optimization or even to provide a performance target for particle engineering efforts thereby reducing the reliance on complex and expensive secondary processing or formulation steps to overcome physical property limitations in raw materials. Furthermore, successful implementation of such validated data-driven models in a user-friendly form would facilitate the transition to Industry 4.0 in pharmaceutical development and manufacture.

References

1. Yang Y, Ye Z, Su Y, Zhao Q, Li X, Ouyang D. Deep learning for in vitro prediction of pharmaceutical formulations. Acta pharmaceutica sinica B. 2019;9(1):177-85.
2. Leane M, Pitt K, Reynolds G, Group MCSW. A proposal for a drug product Manufacturing Classification System (MCS) for oral solid dosage forms. Pharmaceutical development and technology. 2015;20(1):12-21.
3. Ticehurst MD, Marziano I. Integration of active pharmaceutical ingredient solid form selection and particle engineering into drug product design. Journal of Pharmacy and Pharmacology. 2015;67(6):782-802.
4. Leane M, Pitt K, Reynolds GK, Dawson N, Ziegler I, Szepes A, et al. Manufacturing classification system in the real world: factors influencing manufacturing process choices for filed commercial oral solid dosage formulations, case studies from industry and considerations for continuous processing. Pharmaceutical development and technology. 2018;23(10):964-77.
5. Abdi H, Williams LJ. Principal component analysis. Wiley interdisciplinary reviews: computational statistics. 2010;2(4):433-59.
6. Education IC. Unsupervised Learning IBM Cloud Learn Hub IBM; 2020 [updated 21/09/2020. Available from: https://www.ibm.com/cloud/learn/unsupervised-learning#:~:text=Unsupervised%20learning%2C%20also%20known%20as,the%20need%20for%20human%20intervention.
7. Li H, Liu Z. Multivariate time series clustering based on complex network. Pattern Recognition. 2021;115:107919.

700c Impact of Material Attributes on Continuous Mixing Quality: A DEM Study

Peter Toson, Marko Matic, Johan Remmelgas, Dalibor Jajcevic, Jakob Rehrl, Michela Beretta, Thomas O'Connor, Abdollah Koolivand, Geng Tian, Scott M. Krull, Johannes G. Khinast

Digital twins are virtual models designed to accurately reflect the physical objects or processes. In pharmaceutical manufacturing applications, digital twins can aid the design of pharmaceutical production processes, which is especially beneficial when the material availability is low and experimentation at production scale is too costly or even impossible. DEM based digital twins offer detailed insight to the process, but require careful calibration of the material behavior. The digital twin is in principle only valid for the calibrated materials. Thus, a sensitivity study is necessary to quantify how robust or sensitive the digital twin is to changes in the material behavior.

The current work was based on a calibrated and validated DEM model of a cohesive powder in a horizontal continuous mixer. The first part of sensitivity study analyses the influence of the individual DEM contact parameters on the residence time distribution (RTD) of the continuous mixer. The aim of this study is to identify the DEM parameters that have the highest impact on the process outcome and thus require particular care during the calibration procedure.

The second part of the sensitivity study analyses the impact of batch-to-batch variation with a concrete example: Two blends with the same formulation but materials from different batches were experimentally characterized and subsequently used to develop a calibrated DEM model. The technical difference to the first part is that all contact parameters are changed simultaneously to produce suitable DEM contact models for both material batches. The DEM results show that different operating conditions react differently to the batch change. Experimental trials have been performed at nine operating conditions with one material batch, and only one of the two DEM parameter sets match the RTDs of all nine operating conditions at once. These results demonstrate that the presented calibration procedure creates a digital twin that is sufficiently sensitive to detect batch-to-batch variability.

Disclaimer: This abstract reflects the views of the authors and should not be constructed to represent FDA's views or policies.

700d *Agglomeration of Fine APIs As a Material Sparing Screening Indicator for Powder Processability*

Sangah Kim, Mirna Cheikhali, Rajesh Dave

Most of the drugs already marketed or under development are poorly soluble, requiring milling to increase the available surface area for dissolution[1-5]. Unfortunately, milling increases interparticle adhesion force or cohesion, arising from van der Waals (vdW) attraction, concerning individual particle's weight [6-8]. This adversely impacts bulk powder properties and hence poor flow and packing and induces large agglomerate formation while delaying dissolution and unfavorably impacting API blend uniformity[9-15]. Excessive cohesion of fine particles could be mitigated through the dry coating of fine powders with nanoparticles such as silica, reducing the extent of agglomeration[2,16]. Reduction in the agglomeration has been reported to improve dissolution rate and API blend uniformity [13, 14], and overall powder processability, including flowability and packing enabling direct compaction [17-19].

Fine API agglomeration, while expected, has not yet been addressed well in the literature. Our recent work has indicated that the agglomerate ratio (mean agglomerate size to mean primary particle size) could provide early indications of inadequate fine powder processability, i.e., flowability, bulk density, dispersibility, and consequently blend uniformity and dissolution. More importantly, our work has also shown that dry coating could significantly help in reducing the powder cohesion, hence reducing agglomeration and enhancing the powder processability. Consequently, a simple assessment of the agglomerate ratio (AR) with and without dry coating could provide a practical guideline during the early formulation development efforts to mitigate the adverse impact of high cohesion on powder processability. Towards that objective, the current work examined the causal relationship between the interparticle cohesion and the bulk powder properties through a particle scale dimensionless parameter, granular Bond number (Bo_g), towards developing a predictive guideline at a reduced need for experimental testing [12, 20].

The dry coating of the model powders (BCS II drugs, three sizes of ibuprofen (d_{50} ~ 5, 10, 20 μm), fenofibrate (d_{50} ~10 μm), and griseofulvin (d_{50} ~10 μm)) was done employing a high-energy vibratory mixer, LabRAM. The examined material properties include primary and agglomerate size distribution using a compressed-air-dispersion-based laser diffraction particle sizer and a gravity dispersion basis dynamic imaging particle sizer, respectively. USP IV method with 12 mM sodium dodecyl sulfate aqueous solution as the dissolution medium was utilized to evaluate dissolution rate, while FT4 was used to assess flowability (FFC) and bulk density (BD). The surface energy of the uncoated and dry coated powders was measured with inverse gas chromatography, iGC-SEA, which operated under the helium environment. Then, normalized agglomerate size was studied as the critical material attribute, and its relationship to Bo_g, FFC, BD, and dissolution rates was found.

As a novel outcome, the prominent role of agglomeration was demonstrated through demonstrating enhanced dissolution of fine poorly water-soluble APIs even when the dry coating was done with hydrophobic silica, R972P. Additional important outcomes include: (1) For all five model powders, dry coating led to enhanced processability. Flowability was enhanced by up to four regimes, while bulk density increased by up to 100 %. (2) The dissolution, expressed as the area under the dissolution curve (AUC) improved by up to 60% due to a 1–2 orders of magnitude reduction in agglomeration.

(3) As another major novelty, agglomeration (AR), quicker material sparing measure, was a key indicator of both processability and dissolution of fine APIs. (4) AR and Bo_g exhibited a power-law relationship for fine APIs, removing the explicit need for assessing Bo_g. (5) AR was well-correlated with both FFC and BD, greatly facilitating the determination of dry coating efficacy for processability enhancement and setting AR ≤ 5 as a bar for powder processability. (6) The natural surface roughness of the uncoated particles was shown to play a significant role in bulk properties, and the prevalent recommendation of 200 nm (or size-based estimation[21]) natural surface roughness is inappropriate. (7) For APIs such as ibuprofen, milling would significantly increase surface energy values, further increasing effective powder cohesion, requiring dry coating-based passivation [8].

References

1. Hu, J.; Rogers, T. L.; Brown, J.; Young, T.; Johnston, K. P.; Williams Iii, R. O., Improvement of dissolution rates of poorly water soluble APIs using novel spray freezing into liquid technology. *Pharmaceutical Research* **2002,** *19* (9), 1278-1284.

2. Han, X.; Ghoroi, C.; To, D.; Chen, Y.; Davé, R., Simultaneous micronization and surface modification for improvement of flow and dissolution of drug particles. *International Journal of Pharmaceutics* **2011,** *415* (1-2), 185-195.

3. da Costa, M. A.; Seiceira, R. C.; Rodrigues, C. R.; Hoffmeister, C. R. D.; Cabral, L. M.; Rocha, H. V. A., Efavirenz dissolution enhancement I: Co-micronization. *Pharmaceutics* **2013,** *5* (1), 1-22.

4. Bhakay, A.; Azad, M.; Bilgili, E.; Dave, R., Redispersible fast dissolving nanocomposite microparticles of poorly water-soluble drugs. *International Journal of Pharmaceutics* **2014,** *461* (1-2), 367-379.

5. Krull, S. M.; Susarla, R.; Afolabi, A.; Li, M.; Ying, Y.; Iqbal, Z.; Bilgili, E.; Davé, R. N., Polymer strip films as a robust, surfactant-free platform for delivery of BCS Class II drug nanoparticles. *International Journal of Pharmaceutics* **2015,** *489* (1-2), 45-57.

6. Nase, S. T.; Vargas, W. L.; Abatan, A. A.; McCarthy, J. J., Discrete characterization tools for cohesive granular material. *Powder Technology* **2001,** *116* (2-3), 214-223.

7. Feng, J. Q.; Hays, D. A., Relative importance of electrostatic forces on powder particles. *Powder Technology* **2003,** *135-136*, 65-75.

8. Han, X.; Jallo, L.; To, D.; Ghoroi, C.; Davé, R., Passivation of high-surface-energy sites of milled ibuprofen crystals via dry coating for reduced cohesion and improved flowability. *Journal of Pharmaceutical Sciences* **2013b,** *102* (7), 2282-2296.

9. Castellanos, A., The relationship between attractive interparticle forces and bulk behaviour in dry and uncharged fine powders. *Advances in Physics* **2005,** *54* (4), 263-376.

10. Yang, J.; Sliva, A.; Banerjee, A.; Dave, R. N.; Pfeffer, R., Dry particle coating for improving the flowability of cohesive powders. *Powder Technology* **2005,** *158* (1-3), 21-33.

11. Chen, Y.; Yang, J.; Dave, R. N.; Pfeffer, R., Fluidization of coated group C powders. *AIChE Journal* **2008,** *54* (1), 104-121.

12. Capece, M.; Huang, Z.; To, D.; Aloia, M.; Muchira, C.; Davé, R. N.; Yu, A. B., Prediction of

porosity from particle scale interactions: Surface modification of fine cohesive powders. *Powder Technology* **2014,** *254*, 103-113.

13. Huang, Z.; Xiong, W.; Kunnath, K.; Bhaumik, S.; Davé, R. N., Improving blend content uniformity via dry particle coating of micronized drug powders. *European Journal of Pharmaceutical Sciences* **2017,** *104*, 344-355.

14. Kim, S.; Bilgili, E.; Davé, R. N., Impact of altered hydrophobicity and reduced agglomeration on dissolution of micronized poorly water-soluble drug powders after dry coating. *International Journal of Pharmaceutics* **2021,** *606*.

15. De Villiers, M. M., Influence of agglomeration of cohesive particles on the dissolution behaviour of furosemide powder. *International Journal of Pharmaceutics* **1996,** *136* (1-2), 175-179.

16. Han, X.; Ghoroi, C.; Davé, R., Dry coating of micronized API powders for improved dissolution of directly compacted tablets with high drug loading. *International Journal of Pharmaceutics* **2013,** *442* (1-2), 74-85.

17. Huang, Z.; Scicolone, J. V.; Han, X.; Davé, R. N., Improved blend and tablet properties of fine pharmaceutical powders via dry particle coating. *International Journal of Pharmaceutics* **2015,** *478* (2), 447-455.

18. Chen, L.; Liu, J.; Zhang, Y.; Zhang, G.; Kang, Y.; Chen, A.; Feng, X.; Shao, L., The toxicity of silica nanoparticles to the immune system. *Nanomedicine* **2018,** *13* (15), 1939-1962.

19. Chen, L.; He, Z.; Kunnath, K.; Zheng, K.; Kim, S.; Davé, R. N., Fine grade engineered microcrystalline cellulose excipients for direct compaction: Assessing suitability of different dry coating processes. *European Journal of Pharmaceutical Sciences* **2020,** *151*.

20. Kunnath, K.; Chen, L.; Zheng, K.; Davé, R. N., Assessing predictability of packing porosity and bulk density enhancements after dry coating of pharmaceutical powders. *Powder Technology* **2021,** *377*, 709-722.

21. Yu, A. B.; Feng, C. L.; Zou, R. P.; Yang, R. Y., On the relationship between porosity and interparticle forces. *Powder Technology* **2003,** *130* (1-3), 70-76.

700e Engineering of Spherical Lactose and D-Mannitol: Improving Excipient Particles for the Optimized Manufacture of Pharmaceutical Solid Dosage Forms

Joana Pinto, Sarah Zellnitz-Neugebauer, Christina Winter, Amrit Paudel

Introduction

Saccharides are naturally occurring compounds widely used in the pharmaceutical industry as excipients. Excipients are added to the formulation alongside the active pharmaceutical ingredient (API) to improve its processability, bioavailability, and stability. Lactose (LAC), a disaccharide, and D-mannitol (MAN), a polyol, are two widely used saccharides in pharmaceutical formulations. These are commonly used as fillers/diluents in solid dosage forms, such as tablets, pellets, capsules, dry powder inhalers (DPIs), etc., and have a critical role in guaranteeing the adequate performance of the API from manufacturing to its delivery to patients. However, to ensure that these excipients adequately fulfill their role in assuring that different dosage forms present the necessary quality attributes, their particle properties have, many times, to be optimized. Spherical particles, compared to more irregular-shaped ones, have some advantages, particularly in terms of processability. Their improved flow can significantly ease the manufacture of solid dosage forms and sometimes even beneficially impact delivery (i.e., DPIs). Likewise, in this work, we compare the ability of different engineering techniques to obtain stable spherical particles of LAC and MAN that could optimize manufacture and yield improved solid dosage forms.

Materials and Methods

Spray-drying and spray-congealing

Spray-drying and spray-congealing were performed using a 4-M8-TriX spray dryer/chiller (ProCepT, Belgium). For spray-drying, the solutions were fed at a low flow rate (1.0 — 2.0 g/min) into a bi-fluid nozzle (Ø 1.2 mm) at a 0.1 — 0.5 bar pressure. The jet was dispersed into a 0.05 m^3 drying tower working at a drying airflow rate of 0.2 — 0.3 m^3/min, and the temperatures were varied between 180.0°C and 100.0°C. The particles were collected at the outlet between 30.0 and 60.0°C, depending on the inlet temperature and airflow rate. For spray-congealing, the materials were first placed in an oven, and once a melt was obtained, this was transferred to a heated vessel kept at 200.0°C. The molten fluid was pumped at 20.0 g/min from the vessel to a heated bi-fluid nozzle (Ø 1.2 mm) and atomized (0.44 bar) into a glass tower chilled at -10.0°C, using a nitrogen flow of 0.60 m^3/min. The solidified particles were collected in a glass vessel at 13.0°C.

Inkjet printing

Inkjet printing was performed using the Randamp;D inkjet system PixDro LP50 (Meyer Burger, Netherlands) at a constant spotting frequency of 1000 Hz. The substrate (PTFE K400) temperature was kept at 50 °C during printing. To obtain particles, ten droplets were ejected onto the same spot and subsequently left to dry for 1 s. This was repeated 100 times, amounting to 100 particles with a dispensed volume of approximately 1000 pL per particle.

Wide-angle X-ray scattering (WAXS) and FT-RAMAN

For WAXS, the powders were filled into 2 mm glass capillaries, and the measurements were performed in the angular range of 17 and 27° 2θ during 600 s at 30 counts/s. For this, monochromatic CuKα radiation (λ = 1.54 Å) in a combined laboratory small and wide-angle camera (S3-MICRO, Hocus X ray Systems GmbH, Austria) was used. The FT-RAMAN measurements were performed in a Raman Station™ 400F (Perkin Elmer, USA) using three exposures during 10 s in seven analysis spots, at a 2 cm^{-1} between 200 — 3278 cm^{-1}.

Scanning electron microscopy (SEM)

The particles prepared were carefully placed on the pin mount specimen holder of the SEM equipment (Zeiss Ultra 55, Zeiss, Germany). Next, the samples were sputtered with platinum and carbon or gold-palladium and were visualized using a detector operated at an acceleration voltage of 5 kV or 22 kV.

Aerodynamic performance

The aerodynamic performance was evaluated via shot weight consistency and next generation impactor (NGI) studies. The shot weight consistency was assessed by shooting the inhalers ten times into a DUSA (Dosage Unit Sampling Apparatus). High-performance liquid chromatography (HPLC) was used to quantify the quantity of surrogate API released. Depending on the inhaler used, the NGI experiments were carried out at 90 or 60 L/min, and the flow was applied for 2 or 4 s, respectively, to ensure that 4 L of air were drawn via the mouthpiece over the inhaler. The API content in each part of the impactor was quantified via HPLC.

Dissolution performance

The dissolution was investigated using a 24- well plate (Greiner Bio-One™ 24-Well-Platten, Germany), where 2 mL of PBS (pH 7.4) preheated at 37°C were added to each well. The well-plate was placed in an incubator shaker adjusted to 250 pm and 37 °C. UV-Vis was used to quantify the surrogate API release after 1, 5, 10, 15, 30, and 60 min.

Results and Discussion

Engineering of LAC and MAN particles

Crystalline particles are preferred over amorphous ones, as the latter have limited stability. Likewise, some of the spray-dried formulations of the sugars were processed in combination with poly(ethylene glycol) 200 (PEG) as this is a known crystal inducer. This was particularly important when engineering LAC, as this is known to be amorphous after spray-drying. MAN was spray-dried alone in the presence of the pore former (PF), ammonium bicarbonate, and PF combined with PEG. Moreover, because MAN does not degrade upon melting and has a less complex polymorphism, this was chosen to be further engineered via spray-congealing and inkjet printing.

Solid-state of the Engineered Particles

The LAC engineered in the presence of PEG showed to be crystalline and a mixture of the α- and β-polymorphs. In turn, MAN spray-dried alone yielded crystalline particles predominantly of the β-polymorph with traces of the α-form. The addition of PF alone and PF + PEG resulted in crystalline particles of MAN composed mainly of the β- and α-polymorphs, respectively. Printing the same MAN formulations resulted in a similar polymorphism for MAN alone and MAN + PF + PEG formulations. However, after printing, the formulations of MAN + PF, although crystalline, were constituted mainly by the α-polymorph. Spray-congealing of MAN resulted in crystalline particles predominately composed of the α-polymorph.

Morphology of the Engineered Particles

All the produced particles were spherical except the spray-dried MAN + PF + PEG, where large chunks of powder were obtained instead of discrete particles (Figure 1). The LAC + PEG particles presented a mean particle size of 60 μm, were monodispersed, and showed a rough needle-like surface. A similar surface structure was also found after printing when the MAN + PF + PEG formulation was processed. The MAN spray-dried alone was constituted of spherical, smooth agglomerated particles. The addition of PF to the formulation of MAN resulted in the production of shattered particles during spray-drying.
In contrast, discreet solid particles with pores were formed for inkjet printing. Moreover, it was observed that all the printed MAN particles had about 100 μm mean particle size and were very monodispersed. In

contrast, the spray-dried MAN particles were much smaller and very polydispersed. Spray-congealing resulted in smooth polydispersed MAN particles with about 100 μm mean particle size.

Performance of the engineered particles

The spherical particles of LAC + PEG were compared with the classical tomahawk LAC monohydrate concerning their potential as DPI carriers. However, when the LAC + PEG particles were tested in a capsule-based device, these did not perform better than their tomahawk counterparts. Compared to other carrier particles, the spray-congealed MAN showed superior performance when used in binary blends with the API in combination with a reservoir-based device. For dissolution, we compared the printed particles of MAN with the spray-dried ones. Generally, a similar dissolution behavior was observed between printed and spray-dried particles, except for the formulations containing MAN + PEG + PF, where a more rapid dissolution was observed for the printed particles.

Conclusions

Our work showed that it is possible to spray-dry crystalline, monodispersed, spherical particles of LAC with a size suitable for use in DPI when adequately engineering the formulation and process parameters. However, if the compound does not degrade upon melting, spray-congealing might be a more suitable technique to manufacture superior inhalation carriers, as shown for MAN. Engineering of MAN via inkjet printing showed to be a potential alternative to spray-drying when the aim is to yield larger, monodisperse particles of saccharide with optimized dissolution behavior.

References

Alwossabi, E.S. Elamin, E.M.M. Ahmed, M. Abdelrahman, J. Medicinal Aromat. Plants **10** (7), 397 (2021). DOI: 10.35248/2167-0412.21.10.397

J.T. Pinto, S. Zellnitz, T.Guidi, F. Schiaretti, H. Schroettner and A. Paudel, Pharm. Res. **38,** 1107-1123 (2021). DOI: 10.1007/s11095-021-03061-5

J.T. Pinto, S. Zellnitz, T. Guidi, E. Roblegg and A. Paudel, Mol. Pharm. **528** (7), 2827-2839 (2018). DOI: 10.1021/acs.molpharmaceut.8b00333

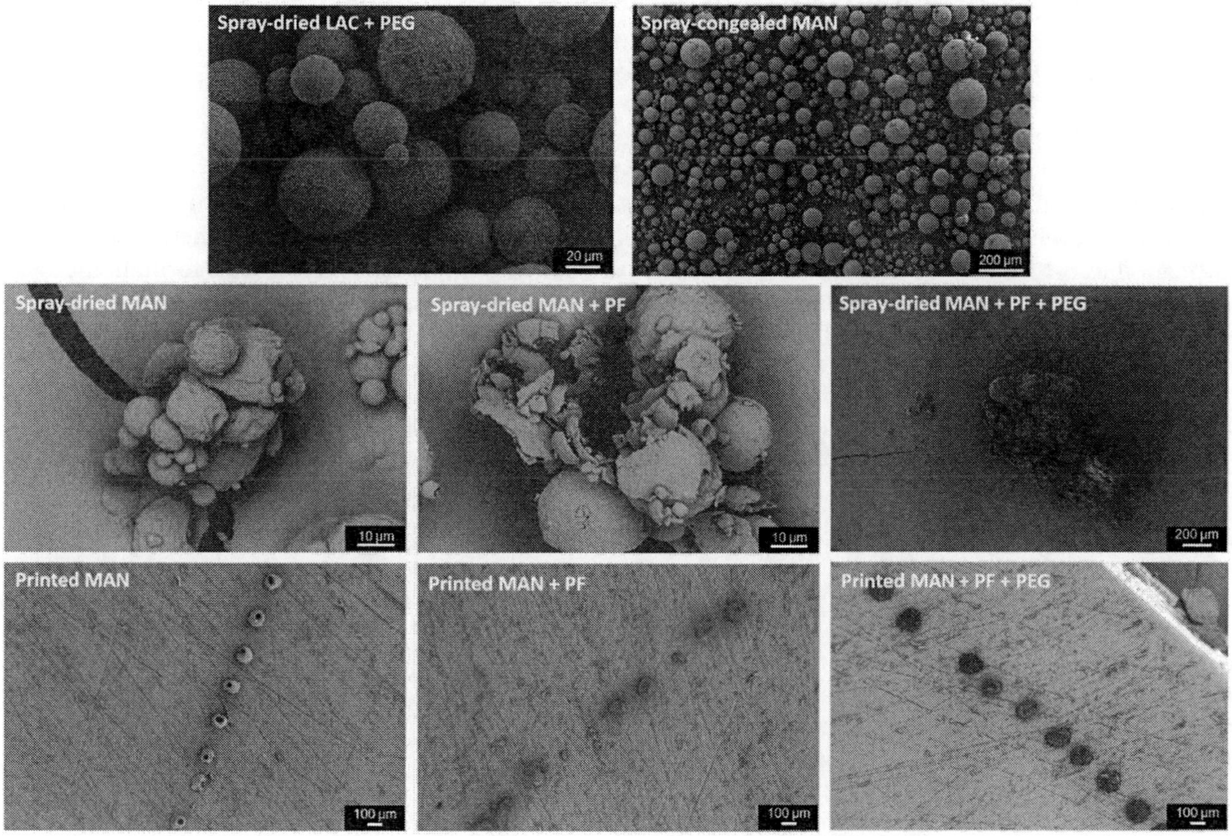

Figure 1: SEM picture of the engineered LAC and MAN particles.

700f Developing a Roadmap to Effectively Spray Dry Biomolecules.

Laura Foley, Ahmad Ziaee, Alice Parkes, Emmet O'Reilly

To date most of the biological therapeutics are delivered in either one of the two ways, via intravenous infusions or subcutaneous injections. However, these are no longer the preferred method of administration for these drugs as they are inconvenient and uncomfortable for some patients while limiting compliance of drug uptake. Moreover, with the current increase in demand for biologics on the market, there is a significant interest for finding innovative formulation and delivery approaches. However, formulating biologics brings many challenges as they are more complex and larger in size than the smaller molecule drugs. One of the major challenges of formulating the biological therapeutics is their increased risk of instability upon exposure to elevated temperatures or mechanical forces. One of the most common methods of increasing the stability of biologics is dehydrating the liquid formulations leading to improved long-term stability at increased temperatures and humidity due to lower molecular mobility and less intermolecular interactions. Freeze drying has been the most frequently used technique for solidification of biologics to date however, high energy consumption, large capital investment for infrastructure and long drying times have made it difficult to cope with the increased market demands.

On the other hand, spray drying is a one-step continuous process to produce dry powder from either a liquid or slurry through rapid drying using an inert gas at elevated temperatures. The spray drying process can potentially lead to particles most suitable for pulmonary delivery application (1-5 μm). The development of a spray drying process of biomolecules has been proven, to be challenging due to their sensitive nature. Additionally, these biomolecules are exposed to relatively high temperatures and atomisation shear forces that could lead to biomolecules denaturation.

This study was carried out on drying of lysozyme as a model biomolecule to investigate the effects of different atomising gas flow rates on its enzymatic activity and physicochemical properties. All samples were spray dried at 50°C outlet temperature, 1.5 ml/min feed rate and 473 L/hr, 601 L/hr and 742 L/hr atomisation gas flow rates; the inlet temperature increased slightly with an increase in atomisation gas flow rate. The applied shear stress during the atomisation process is a significant factor in determining the final stability of biomolecule that has been overlooked in previous research activities carried out on lysozyme.

The effect of atomisation gas flow rate on the residual bioactivity of lysozyme after spray drying can be seen in Figure 1. It has been shown that lysozyme in the solid form is increasingly deactivated as the atomisation gas flow rate increases. The lowest residual bioactivity of lysozyme was observed in samples sprayed at the highest atomisation gas flow rate (742 L/hr). On the other hand, samples atomised at the lowest atomisation gas flow rate (473 L/hr) showed the highest residual bioactivity (88%). These experiments show the direct implication of atomisation gas flow rate *i.e.* shear force on the denaturation of lysozyme. In addition, spray drying of lysozyme at various atomisation gas flow rates leads to the production of powders with particle sizes in the range of 4μm to 16 μm. Particle size distribution graphs show that powder C produced using higher atomisation gas flow rates lead to the smallest particle size of 4 μm and D50 of 8 μm, which suits best for pulmonary delivery applications. However, the low bioactivity of lysozyme at this level of atomisation gas flow rate due to high applied shear force needs to be addressed. The next phase of this project will be focused on addressing the stability challenge of biomolecules at high atomisation gas flow rates and achieving high proportion of fine particles (1-5 μm). These results have implications in changing the current biologics delivery methods *i.e.*, intravenous, and subcutaneous injections with inhalable drug delivery methods. This could promote better patient compliance along with faster delivery outcomes.

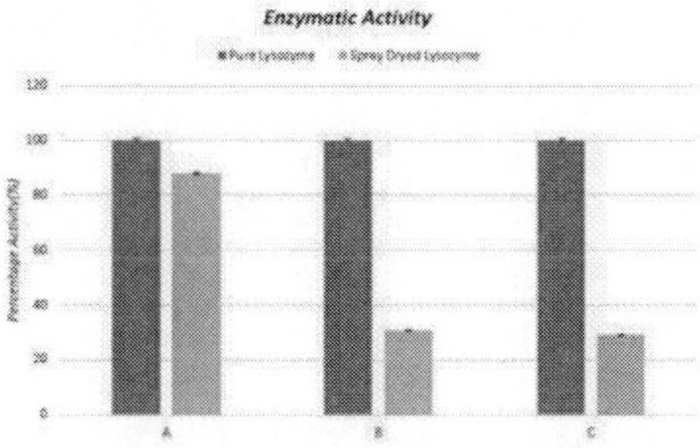

Figure 1: (A) The atomisation flow rate is set at 40, (B) the atomisation flow rate is set at 50 and (C) the atomisation flow rate is set at 60. This Graph shows the percentage of enzymatic activity after spray drying lysozyme against the enzymatic activity before processing

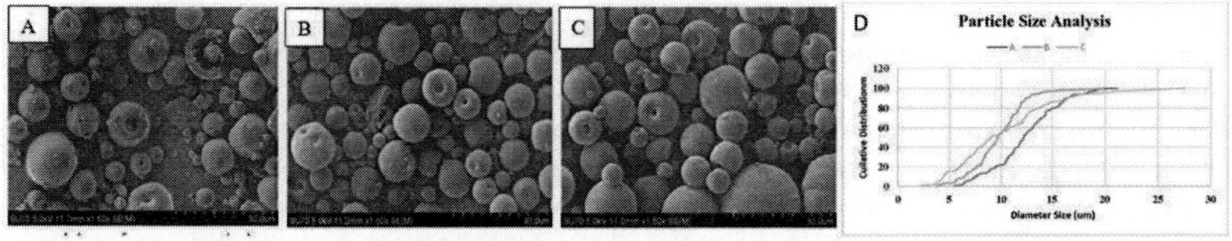

Figure 2: Scanning Electron Microscopy of spray dried lysozyme at the various atomisation flow rates. (A) The atomisation flow rate is set at 473 L/hr, (B) the atomisation flow rate is set at 601 L/hr, (C) the atomisation flow rate is set at 472 L/hrand (D) graphical distribution of the particle size of A,B and C respectively

700g Lipid-Based Particle Engineering Via Spray-Drying for Drug Targeting to the Lung

Đana Crvenjak, Carolina Corzo, Sharareh Salar-Behzadi, Sebastian Reyer, Dirk Lochmann, Andreas Zimmer

Motivation and Scope

Rifampicin (RIF) is an antibiotic listed as an essential medicine by the World Health Organization. It is administered orally or intravenously in the therapy against Tuberculosis. However, RIF, like other antibiotics, can cause serious adverse effects, such as hepatotoxicity. Delivery of RIF directly to the site of action via inhalation can reduce the risk of such effects by diminishing systemic exposure. Furthermore, pulmonary delivery of RIF can lead to dose reduction and less frequent administration [1]. For that, RIF-loaded particles that are able to reach deep lung deposition and trigger phagocytosis of alveolar macrophages (AM) are required. These requirements can be provided by loading RIF into lipid-based excipients (LBE) followed by particle engineering via spray-drying. Targeted particle attributes are given by the interplay of lipid's material attributes and process parameters. In this work, selected LBE were used to engineer particles for pulmonary delivery of RIF to be administered as dry powder for inhalation (DPI).

Materials

Diglycerol full ester of behenic acid (PG2-C22f) commercially available as Witepsol® PMF222 was selected as LBE due to its superior processability via spray-drying. As previously shown by our group in [2], the high melting point (Tm=72.5°C), monophasic crystallization and absence of polymorphism of PG2-C22f lead to high process yield. Glyceryl stearate citrate (Imwitor®372P) was used as lipid-based emulsifier, selected due to the similar solid state structure to PG2-C22f and high HLB (10-12).

Methods

Setting critical particle attributes: Inhalable particles with mass median aerodynamic diameter (MMAD) between 1-5μm were required for suitable lung deposition. After deposition, activation and uptake by AM were envisaged. Therefore, particles of volume mean diameter (VMD) below 4μm (optimal 2.1 μm) and negative surface charge, measured via Zeta Potential (ZP) [3] were designed.

Spray-drying: An aqueous-based suspension was prepared for spray-drying. PG2-C22f and Imwitor372P (1:1 mass ratio) were melted at 80°C. RIF was dissolved in the molten mixture (2.5% over lipid mass). Pre-heated water was added to the mixture to provide a solid content of 0.5% w/w. High-shear stirring was applied. The mixture was left to cool down at room temperature to allow the formation of solid lipid-particles loaded with RIF within a suspension. The suspension was atomized through a bi-fluid nozzle of 0.4mm (1.14 bar) at 2 g/min into the spray-drying chamber at inlet temperature (Tin) of 75°C. Particles were recovered via cyclone.

Particle characterization: The morphology of particles was observed via Scanning Electron Microscopy (SEM). Particle size distribution was analyzed via light diffraction. Bulk and tapped densities were used to assess Hausner ratio. Solid state and stability were evaluated via differential scanning calorimetry (DSC) and small and wide angle X-ray scattering (SWAXS) on freshly prepared samples and after storage for 3 months at 25°C/60%r.H (LTC) and 40°C/75%r.H (AC). The aerodynamic performance was assessed via the next generation impactor (NGI) coupled to Aerolizer dry powder inhaler. Six hypromellose capsules were loaded with 30mg of powder. An inhalation flow of 60 l/min for 4 seconds was used. MMAD and in vitro deposition were obtained. Particle uptake was assessed in THP-1 cells by exposing them to suspended particles for 24 hours, followed by drug extraction and quantification via HPLC.

Results

Physical characterization

Spray-drying of RIF-loaded lipid-suspension resulted in high yield of particles (83%). This is in agreement with our previous findings [2]. SEM images evidenced spherical particles of slightly rough surface. No internal voids were formed during spray-drying due to slow evaporative conditions ($T_{in}<T_{b,water}$). A suitable VMD of 2.41 ± 0.07μm was provided. The flowability of particles was characterized by a Hausner ratio of 1.90, associated to high powder cohesiveness. Negative surface charge (ZP = -36.25 ± 7.97 mV) was found in the particles, which is attributed to the carboxyl groups within the structure of PG2-C22f and glyceryl stearate citrate.

Solid state

The solid state of particles and stability thereof was confirmed via DSC and SWAXS. No alterations of endothermic events were observed. Endothermic transitions of RIF were not found, depicting the solubilization of the drug within the lipid matrix. SWAXS analysis confirmed the absence of crystalline RIF signals (Fig. 3). The crystalline structure of the lipid matrix corresponds to a stable α-form of both excipients, arranged in a mixed lamellar structure of 68A. After storage, however, slight inclination of lamellae due to potential appearance of the β-form of Glyceryl stearate citrate was observed.

Aerodynamic and cellular performance

The aerodynamic performance of the particles as lipid-based DPI showed that, in spite of the high cohesivity of the particles, they were successfully aerosolized. A high emitted fraction of 94.8 ± 0.2% was found and associated to the lipidic particle surface. Suitable MMAD of 2.44 ± 0.34μm depicted the readily inhalability of the particles with high FPF of 61.2 ± 0.1%. Exposure of THP-1, model cell line of AM, to the particles showed 88.8 ng of RIF/Mio cell taken up; whereas 50.5 ng /Mio cell were taken up from exposition to a RIF-solution.

Conclusion

Lipid-microparticles loaded with RIF were successfully prepared through an aqueous-based spray-drying process of high yield. The particles were produced through an exclusively lipid-based formulation using diglycerol full ester of behenic acid and glyceryl stearate citrate. Optimal particle attributes and outstanding aerodynamic performance as lipid-based DPI were achieved. Such attributes resulted in the targeted delivery of RIF to AM, which in turn offers a vast potential for dose reduction and reduced adverse side effects.

Acknowledgment: IOI Oleo GmbH, the Austrian Funding Agency (FFG) and the Doctoral Academy NanoGraz.

640a Intelligent High-Throughput Intervention Testing Platform in *Daphnia*

Yongmin Cho, Marc Kirschner, Leonid Peshkin

Developing pharmacological interventions that slow down the aging process and consequently postpone the onset and progression of age-associated diseases is highly sought after. However, since the mechanisms driving the aging process are not well understood, there currently exist few druggable targets for anti-aging treatments and therefore the evaluation of drug effects on the aging process requires the development of new high-throughput screening platforms. A key component of new high-throughput screening platforms will be quantitative biomarkers of aging. Since individuals may not age at the same rate, quantitative biomarkers of aging are valuable tools to measure physiological age, assess the extent of healthy aging, and potentially predict not only health and lifespan but also age-related outcomes for individuals within a population, even at the early age. Molecular biomarkers (often based on gene expression) are robust quantitative metrics and can reflect some of the molecular mechanisms underlying the aging process, but often require sacrifice of the subject, laborious sample processing, and constitute a single data endpoint. Phenotypic biomarkers can be harder to quantify but are fairly easy to obtain, non-invasive, and therefore are possible to repeatedly assay over the entire life of the subject and in future generations.

In this study, we present a novel platform for testing the effect of interventions on life- and health-span of a short-lived semi-transparent freshwater organism, sensitive to drugs with complex behavior and physiology – the planktonic crustacean *Daphnia magna.* Within this platform, dozens of complex behavioral features of both routine motion and response to stimuli are continuously accurately quantified for large homogeneous cohorts via an automated phenotyping pipeline. We build predictive machine learning models calibrated using chronological age and extrapolate onto phenotypic age. We further apply the model to estimate the phenotypic age under pharmacological perturbations including metformin and rapamycin. Our platform provides a scalable framework for drug screening and characterization in both life-long and instant assays as illustrated using a long-term dose-response profile of metformin and a short-term assay of well-studied substances such as caffeine and alcohol.

640b Cellular Proton Motive Force: A Therapeutic Target for Eradicating Methicillin-Resistant *Staphylococcus Aureus* Persister Cells

Sreyashi Ghosh, Sayed Golam Mohiuddin, Pouria Kavousi, Mehmet Orman

Persister cells are a small fraction of non-proliferating variants in a cell population that can tolerate lethal doses of antibiotics. These cells can resume their growth upon removal of antibiotics. In this study, we aim to identify therapeutic strategies to eliminate persister cells of Methicillin-resistant *Staphylococcus aureus* (MRSA) strains. MRSA remains one of the main virulent pathogens responsible for serious hospital-acquired infections including bloodstream infections and ventilator-associated pneumonia[1]. Because of their emerging resistance to some of the major classes of antibiotics, it is becoming increasingly difficult to eradicate MRSA persisters. Although previous studies have shown the effectiveness of proton motive force (PMF) inhibitors at eradicating bacterial cell population [2,3], further research is needed to investigate the antipersister potency of PMF inhibitors. Using two different MRSA strains and known PMF inhibitors and conventional antibiotics, we showed that anti-persister potency of the inhibitors depends on their ability to disrupt PMF and permeabilize cell membranes (**Figure 1**). To verify this observed correlation, we screened a small library containing chemicals that can inhibit various electron transport chain complexes and identified several drugs, such as nordihydroguaiaretic acid, gossypol, trifluoperazine, and amitriptyline, that strongly disrupted PMF in MRSA cells by dissipating either the transmembrane electric potential or the proton gradient. These identified inhibitors permeabilized cell membranes and reduced persister levels below the limit of detection. Overall, our study demonstrated that the PMF inhibitors can be highly effective bactericidal drugs with the potential to eradicate persister cells.

References:

1. Enright, M. C. *et al.* The evolutionary history of methicillin-resistant Staphylococcus aureus (MRSA). *Proc Natl Acad Sci U S A* **99**, 7687–7692 (2002).
2. Farha, M. A., Verschoor, C. P., Bowdish, D. andamp; Brown, E. D. Collapsing the proton motive force to identify synergistic combinations against staphylococcus aureus. *Chemistry and Biology* **20**, 1168–1178 (2013).
3. Stokes, J. M. *et al.* A Deep Learning Approach to Antibiotic Discovery. *Cell* **180**, 688-702.e13 (2020).

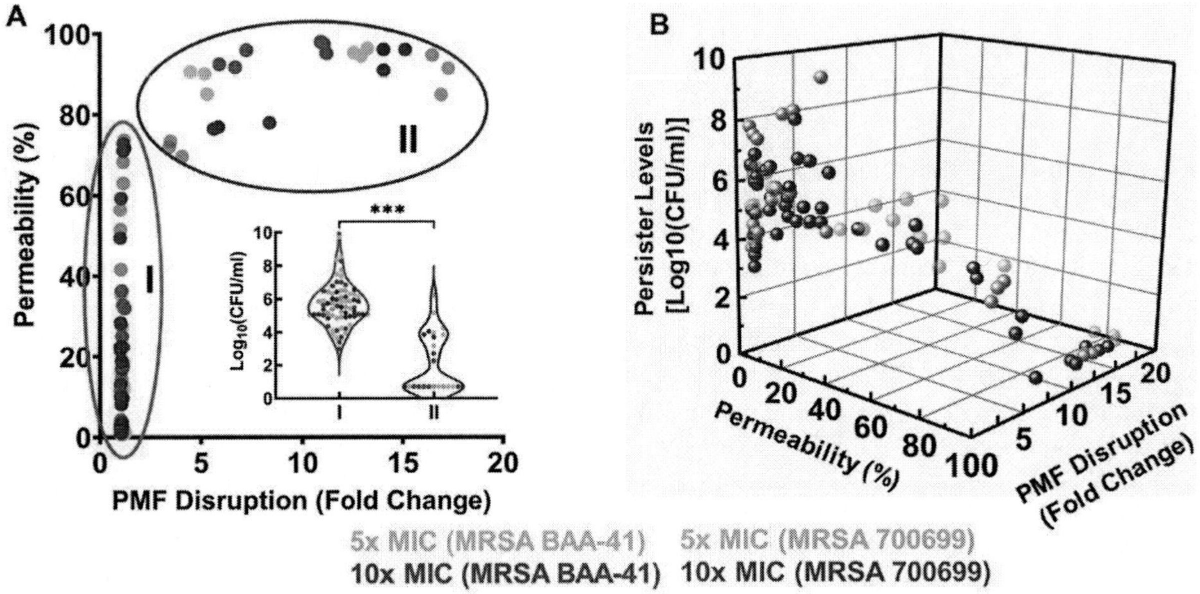

Figure 1: Correlation between PMF disruption, membrane permeability and persister level. MRSA BAA-41 and MRSA 700699 cells were grown to the exponential phase (OD600 of ~0.1) in Mueller–Hinton broth and treated with PMF inhibitors and conventional antibiotics at 5× and 10× of minimum inhibitory concentrations (MIC). After 1 h treatment, cells were stained with PI (20 µM) dye for measuring the percentage of membrane permeabilization. After 6 h of treatment, cells were washed to remove the chemicals, and spotted on Mueller–Hinton agar plates to obtain CFU counts. For determining the extent of PMF disruption, exponential phase cells were transferred to $DiSC_3(5)$ assay buffer and stained with $DiSC_3(5)$. After reaching an equilibrium state (t = 30 minutes), the cells were treated with PMF inhibitors and conventional antibiotics at the indicated concentrations and fluorescence levels were with a plate reader. **(A)** Two-dimensional scatter plots including all data point for PMF inhibitors and conventional antibiotics for all concentrations and strains tested. In panel A, the red circle indicates cluster I and the blue circle indicates cluster II. The persister levels corresponding to each cluster are presented in the inset. A Student'st-test with unequal variance was used to find the statistical significance between the persister levels of clusters I and II (***P < 0.0001). **(B)** Three-dimensional scatter plots including all data points for PMF inhibitors and conventional antibiotics for all concentrations and strains tested.

640c Identification of Apolipoprotein E4 Inhibitors for Alzheimer's Disease Therapy through Large-Scale Virtual Screening

Zuyi Huang, Emily Krass, Tianhua Zhai

Alzheimer's disease (AD), a neurodegenerative disorder, is characterized by its ability to cause memory loss and damage other cognitive functions. In 2015, it was estimated that almost 50 million people were affected by this disease, and this number is expected to increase in coming years as the increasing average life expectancy increases [1]. There are also concerns over the social and economic impact of the disease since the loss of function that affected individuals experience can lead to their dependence on outside care. Since current treatments mainly focus on managing symptoms without directly targeting AD pathology, there is a need to investigate new approaches and treatments for inhibiting potential protein targets involved in AD[2]. The two main pathological hallmarks of AD are the amyloid beta (Aβ) and tau proteins which accumulate in the form of plaques and neurofibrillary tangles, respectively[2]. Aggregation of beta amyloid plaques (Aβ) and neurofibrillary tangles in brain are responsible for development of AD. Nonetheless, many attempts targeting Aβ and tau proteins have been proven mostly unsuccessful[3]. Currently, evidences mounts that targeting Apolipoprotein E ε4 may combat Alzheimer's Pathogenesis[4]. Apolipoprotein E has three different major alleles that are found in humans, leading to different protein isoforms, apoE2, apoE3, and apoE4. APOE ε4 is the strongest genetic risk factor for sporadic Alzheimer's disease[5]. ApoE4 has been shown to increase Aβ accumulation by directly binding to Aβ and stabilizing it as well as interfering with Aβ and tau protein clearance by competing with receptors[5]. In this study, Apolipoprotein E4 (ApoE4) was chosen as the target based on its role in the main pathological hallmarks of AD. The structure and function of ApoE4 protein were investigated by previous studies[6]. The crystal structures can be obtained from Protein Data Bank with code 6NCN and 6NCO. We aimed to discover small-molecule inhibitors binding to ApoE4 protein to provide potential therapeutical options for Alzheimer's disease.

Since there are millions of chemical compounds available in compound databases, such as FDA approved drugs that were listed in the DrugBank database, and experimental compounds that were registered and are available for purchase through ChemBridge, it is nontrivial to conduct experimental screening of each chemical compound against ApoE4. Therefore, a computational approach was then developed to identify small-molecule inhibitors of ApoE4 protein. Ligand-protein docking was conducted large-scale virtually screening. Among the existing ligand-protein docking, such as AutoDock, FlexX, ICM and GOLD, ICM was reported with the best performance[7]. It was able to predict structures for 93% complexes within the acceptable accuracy. Therefore, ICM was used as the docking platform to screen potential small molecule inhibitors for the target protein ApoE4 from large compounds databases. The crystal ligand was redocked in ApoE4 protein and got binding energy score -18.01 kcal/mol. In the protein-ligand docking software, lower scores indicate stronger binding affinity between the compounds and ApoE4. Compounds with binding scores below -20.0 were of special interest in this work, as they were predicted to have better binding affinity than all the crystal ligands, regardless of pocket. We virtually screened 1.5 million compounds from FDA-approved drug library and Chembridge database. By attempting to reposition FDA-approved drugs (approximately 2,500 drugs), resources are saved since identified compounds should already have data related to clinical trials and toxicity and be classified as safe for human use. The ChemBridge database was further used to evaluate approximately 1.5 million compounds binding to ApoE4 protein. The ChemBridge was a much more comprehensive database of drug-like compounds; however, since these compounds are experimental, they are not necessarily FDA-approved or extensively studied.

Subsequently, chemical clustering program select putative inhibitors with diverse common substructures. The rationale for this is that compounds share similar structure may have similar properties. Therefore, the number of ApoE4 inhibitor candidates can be further reduced after compounds are clustered according to their structure similarity. An unweighted pair group method with arithmetic mean (UPGMA) clustering method was conducted to cluster compounds based on substructures. UPGMA, which achieved the best result compared to Butina-based clustering, is an agglomerative, hierarchal approach which is used for drug similarity studies[8]. We further selected inhibitors and extracted common structures from each group. The final candidates were evaluated on the following drug-like properties: partition coefficient (LogP), solubility (LogS), permeability through blood brain barrier (BBB score), predict CACO-2 permeability LogP (CACO2 score), toxicity (LD50).

From the DrugBank screening, there were 15 compounds that had a docking score below -20.0. In addition, several compounds were shown to bind to more than one pocket. Docking results ranged from -24.19 to -20.03 kcal/mol. From the ChemBridge screening, there were 1391 putative strong binders of ApoE4 were identified by docking-based virtual screening with docking score ranged from -27.35 to -20.02 kcal/mol. A key group of compounds that were identified as potentially effective were non-steroidal anti-inflammatory drugs (NSAIDs). Anti-inflammatory effects could be beneficial since neuroinflammation is a contributing factor to AD and its pathology[1]. Numerous epidemiological studies in a range of ethnogeography populations have identified that NSAID use is associated with a lower risk of developing Alzheimer disease, and this association was suggested to be causal by numerous intervention-based studies in animal models[9]. Therefore, our findings suggested an interesting hypothesis that NSAID drugs might act as ApoE4 inhibitors against Alzheimer's disease. The 1391 compounds identified from Chembridge were clustered based on UPGMA clustering method. In total, ten common substructures were identified. Since the compounds clustered in the same group share similar structure, core compounds from each group were extracted as the representative inhibitor. This results in a final hit list that contains 577 compounds. The putative inhibitors were found containing key scaffold/functional groups including benzene sulfonamide, diaromatic rings, pyrazole, thiophene, quinoline.

In conclusion, total 592 potential ApoE4 inhibitors were newly identified through computational approaches, which would provide drug candidates for further experiment assays. Therefore, in vitro testing is needed to confirm binding predictions; however, this could be challenging due to the stability of ApoE4. If in vitro testing were to confirm effective compounds, another potential direction could be to use data about binding and PROTAC technology to design a linker that would allow entire protein degradation rather than just inhibition.

References

1. Loera-Valencia, R. *et al.* Current and emerging avenues for Alzheimer's disease drug targets. *Journal of Internal Medicine* (2019) doi:10.1111/joim.12959.
2. Cummings, J. New approaches to symptomatic treatments for Alzheimer's disease. *Molecular Neurodegeneration* (2021) doi:10.1186/s13024-021-00424-9.
3. Congdon, E. E. andamp; Sigurdsson, E. M. Tau-targeting therapies for Alzheimer disease. *Nature Reviews Neurology* (2018) doi:10.1038/s41582-018-0013-z.
4. Uddin, M. S. *et al.* APOE and Alzheimer's Disease: Evidence Mounts that Targeting APOE4 may Combat Alzheimer's Pathogenesis. *Molecular Neurobiology* (2019) doi:10.1007/s12035-018-1237-z.
5. Serrano-Pozo, A., Das, S. andamp; Hyman, B. T. APOE and Alzheimer's disease: advances in genetics, pathophysiology, and therapeutic approaches. *The Lancet Neurology* (2021)

doi:10.1016/S1474-4422(20)30412-9.

6. Petros, A. M. *et al.* Fragment-Based Discovery of an Apolipoprotein E4 (apoE4) Stabilizer. *J. Med. Chem.* (2019) doi:10.1021/acs.jmedchem.9b00178.
7. Bursulaya, B. D., Totrov, M., Abagyan, R. andamp; Brooks, C. L. Comparative study of several algorithms for flexible ligand docking. *J. Comput. Aided. Mol. Des.* (2003) doi:10.1023/B:JCAM.0000017496.76572.6f.
8. Kruger, F., Fechner, N. andamp; Stiefl, N. Automated Identification of Chemical Series: Classifying like a Medicinal Chemist. *J. Chem. Inf. Model.* (2020) doi:10.1021/acs.jcim.0c00204.
9. Rivers-Auty, J., Mather, A. E., Peters, R., Lawrence, C. B. andamp; Brough, D. Anti-inflammatories in Alzheimer's disease—potential therapy or spurious correlate? *Brain Commun.* (2020) doi:10.1093/braincomms/fcaa109.

640d Polymer Lung Surfactant: A First-in-Class Synthetic Surfactant Therapeutic for Acute Respiratory Distress Syndrome (ARDS)

You-Yeon Won, Daniel Fesenmeier, Seyoung Kim

Acute Respiratory Distress Syndrome (ARDS) is a life-threatening condition affecting about 200,000 patients in the United States each year. When ARDS occurs, the function of native lung surfactant becomes impaired, contributing to a severe decrease in blood oxygenation and eventually multiple organ failures. There are currently no therapeutic surfactant formulations which have been shown to be effective in treating this condition; all prior clinical trials testing animal-derived lung surfactants in ARDS patients were unsuccessful. Our laboratory has been exploring a radically different approach, in which, instead of conventional lipid/protein-based formulations, synthetic biocompatible polymers are used as the active therapeutic ingredient. This investigation led to the development of a treatment which uses an amphiphilic block copolymer formulation. This formulation has been shown to be more effective in ARDS mouse models compared to clinically available animal derived surfactants. The polymer formulation is a promising candidate for lung surfactant replacement therapy as it forms an insoluble monolayer at the alveolar air-water interface which is resistant to serum protein/phospholipase deactivation. This talk will present accounts of our research endeavors over the past few years that enabled the development of this first-in-class polymer lung surfactant therapy for ARDS treatment.

640e An Innovative Therapeutic Fusion Protein Approach for Chronic Wound Healing

Hazim Aljewari, Shadrach Ibinola, Stephanie Beitle, Ahmed Elmasheiti, Josh Sakon, Robert Beitle

In this study, the first steps in designing, expressing and *in vitro* testing were taken to investigate the properties of a potential therapeutic recombinant designed to release basic fibroblast growth factor (bFGF) at a wound site. It was hypothesized that by fusing bFGF to two collagen binding domains separated by a collagenase sensitive linker, the recombinant would preferentially locate to the site of a tear and eventually be released via enzymatic cleavage. This presentation will discuss the first steps in achieving this goal. The novel fusion protein was expressed in *Escherichia coli* BL-21using batch and fed batch techniques and purified by liquid chromatography from the lysate. Results showed that the target fusion protein was expressed and is sensitive to the protease, providing the initial data needed to establish the continued work on the project

640f Peg-Conjugated Extracellular Mega-Hemoglobin for Diverse Oxygen Therapeutic Applications

Chintan Savla

Motivation: The newest generation of hemoglobin-based oxygen carriers (HBOCs) has found generic applications as oxygen therapeutics apart from their traditional use as resuscitative fluids. Recently, HBOCs have been used in organ and graft preservation, *ex-vivo* machine perfusion, cell growth media supplements, hypoxic tumor models, and wound healing. These diverse applications have necessitated the development of oxygen therapeutics that can be used for multiple applications. Steering away from traditional mammalian hemoglobins, this work focuses on the mega-hemoglobin (LtEc) of the annelid *Lumbricus terrestris* for use as an oxygen therapeutic. Its large size (~30 nm) reduces the possibility of *in vivo* extravasation and scavenging of endothelial-derived nitric oxide (NO). The large molecular diameter of LtEc along with its' high structural stability, low rate of auto-oxidation, moderate colloid osmotic pressure (COP), and human blood-like oxygen binding properties make it an ideal candidate for use as a resuscitative fluid in trauma treatment. Subsequently, LtEc is also a promising candidate for use in graft preservation and as an oxygen supplement in cell culture media because of its ability to stay in the reduced ferrous form for long periods of time. However, to mask the exterior protein surface and reduce potential immunogenic activity, surface conjugation is necessary. In this study, we conjugated polyethylene glycol (PEG) to the surface of LtEc to camouflage it and performed comprehensive *in vitro* characterization and preliminary *in vivo* studies to investigate the potential of PEG modified-LtEc as an oxygen therapeutic.

Methods: Purification of LtEc from *Lumbricus terrestris* worms was performed by modifying previously devised protocols from our lab with a focus on improving the purity of the resulting product. In short, Canadian nightcrawlers (*Lumbricus terrestris*) were rinsed to remove dirt and mucus prior to homogenization in a modified Tris buffer. The homogenate was centrifuged via multiple cycles to eliminate worm debris and dirt and obtain a cloudy red supernatant. Post vacuum filtration, multi-step tangential flow filtration through 0.65 μm and 0.22 μm hollow fiber filters was performed to eliminate large aggregates and bacterial components. Continuous volume diafiltrations were carried out over a 500 kDa filter to eliminate a majority of the unwanted impurities. Retentate and permeate samples were analyzed using analytical size exclusion chromatography (SEC) to track loss of proteins and the purification was concluded when the retentate reached >99% purity. *In vitro* biophysical characterization included dynamic light scattering (DLS), oxygen equilibrium analysis, rapid deoxygenation kinetics, gel electrophoresis (SDS-PAGE), and auto-oxidation kinetics. Surface modification of purified LtEc was carried out using thiol-maleimide chemistry by converting surface lysine residues to thiols and attaching PEG chains with monofunctional ends. Unreacted components were removed, and the conjugated products was analyzed using amino acid residue assays, DLS, SEC, and SDS-PAGE to confirm surface modification. Oxygen equilibrium properties and other *in vitro* properties were tested and compared to unmodified material. The degree of surface camouflage was engineered by exploiting the numerous lysine residues on the LtEc surface and desirable *in vitro* properties were obtained.

Results: The modified LtEc purification protocol was successful in obtaining an LtEc product with >99% purity, which was verified using SEC and SDS-PAGE analyses. This was an important result as the removal of non-mammalian proteins and impurities is crucial for a pharmaceutical product to be used in transfusion medicine. Biophysical characterization was performed to determine advantageous biophysical properties of LtEc such as its' large hydrodynamic diameter, moderate oxygen affinity, high oxygen cooperativity, low auto-oxidation rate, and fast deoxygenation kinetics. *In vivo* analysis in a golden Syrian hamster model corroborated that the ultra-pure product performed better in animals than product purified using older protocols with significantly better flow in arterioles and higher functional capillary density. Various degrees of surface modification with PEG resulted in PEG-LtEc products with varying biophysical properties. Even though the hydrodynamic size varied little, we observed changes in molecular masses of subunits using the SEC, SDS-PAGE, and matrix-assisted laser desorption ionization (MALDI) mass spectral analyses, further confirming successful surface camouflage. The attachment of PEG resulted in a product with higher oxygen affinity, lower oxygen cooperativity, higher auto-oxidation rates, and slower kinetic rate of oxygen offloading than LtEc. In spite of the changes to the biophysical properties, it is expected that modified products will yield higher circulatory times and lesser immunogenic responses, which will be tested next.

Takeaways: The streamlined purification protocol devised in the study allows for the production of an extracellular hemoglobin with >99% purity. This purity is essential to enable a wide variety of applications of LtEc as an oxygen therapeutic. We show varying extents of camouflage of the LtEc surface with PEG and their effect on biophysical properties of the product. The products are expected to have longer circulation time *in vivo* owing to its surface masking. Future animal tests will test this hypothesis. However, the use of LtEc can be expanded to cell media supplement and graft preservation, making it a diverse oxygen therapeutic.

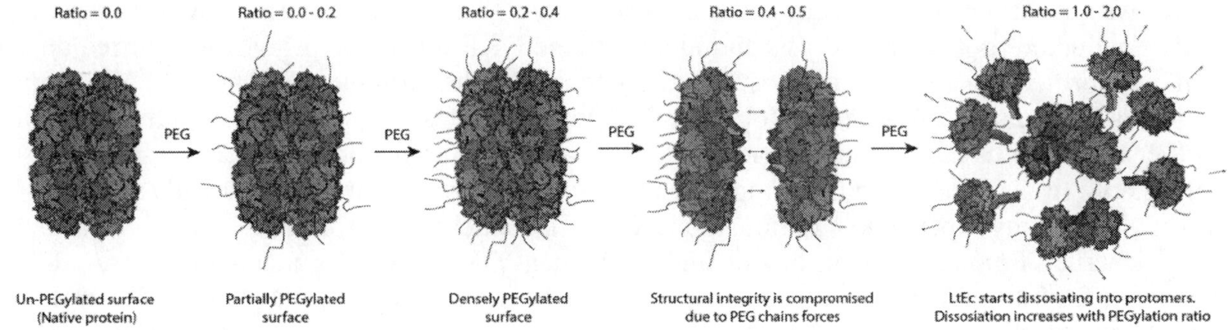

207a Mechanistic Study of Diketopiperazine Formation during Solid Phase Peptide Synthesis of Tirzepatide

Jingyao Wang, Mark Berglund, Fareed Sayyed, Matthew C. Embry, Stephen Groskreutz, Martin Johnson, Timothy M. Braden, Michael Kopach

This study focused on investigating diketopiperazine (DKP) and the formation of associated double amino acid deletion impurities during linear solid phase peptide synthesis (SPPS) of tirzepatide (TZP), a 39 amino acid synthetic peptide. The C terminus desProPro double amino acid deletion impurity was observed at significant levels during the synthesis of TZP intermediate. In this study, we initially conducted root cause analysis and identified the DKP formation primarily occurred during the Fmoc deprotection reaction and post-coupling ageing of the unstable API intermediate. Similar phenomena have also been observed for other TZP intermediates which contain penultimate proline and other C Terminus intermediates for both hybrid and linear synthetic approaches. During post-coupling hold times, it is found that Fmoc- deprotection can proceed auto catalytically in various solvents without any piperidine addition, along with subsequent DKP formation on the deprotected peptide. Detailed Fmoc- self-deprotection and subsequent DKP formation mechanisms are proposed and account for intramolecular pathways triggered by the amine group of the penultimate proline and intermolecular pathways catalysed by the deprotected amine group in the peptide and DKP by product. Reaction kinetics parameters are solved for various peptide API intermediates under different solvents and temperature conditions to describe rate of impurity formation. As general synthesis practice described in past literature, we also found use of oxyma additives and lower storage temperature can greatly stabilize the peptide intermediated to DKP by-products.

207b Strategies for Mitigation and Control of PS80 Degradation in a Monoclonal Antibody Downstream Process

Kevin Chang, Hope McMahon, Zifan Gong, Andreas Sophocleous, Robert Luo, Antonio Ubiera

Polysorbates (PS), including PS80, are surfactants commonly used in the biopharmaceutical industry to prevent adsorption, aggregation, and/or particulate formation in monoclonal antibody (mAb) formulations. In recent years, cell culture process intensification and extensive process development activities have led to PS80 degradation challenges, attributed to the enzymatic activity of trace levels of residual host cell proteins (HCPs). Upon long-term storage of formulated mAb drug substance (DS) at ambient temperature, hydrolysis of PS80, catalyzed by these low levels of residual HCPs, can lead to precipitation and formation of undesired sub-visible and visible particulates, and thus, impact the drug product (DP) manufacturing and product control strategy. One approach to address these risks is to establish an integrated mitigation and control strategy for PS80 degradation in the DS downstream process. This work focuses on mitigating and controlling PS80 degradation through various unit operations in DS downstream process development to support technology transfer from bench-scale to clinical/commercial manufacturing facilities and to maintain formulated mAb stability throughout its shelf life. Specifically, we sought opportunities to mitigate the risk of PS80 degradation by evaluating the ability of Protein A chromatography and depth filtration post virus inactivation (VI) to clear HCPs during the processing of a mAb. Furthermore, we identified that process hold times at various temperatures prior to freezing DS could lead to consistently below-target DS PS80 concentration. This below-target PS80 concentration coupled with PS80 degradation could pose significant product risk. For this reason, a DS PS80 degradation study was conducted at cold (2 – 8 °C) and room (25 °C) temperatures to determine PS80 degradation rate and appropriate hold times during DS manufacturing. As a result, minimal PS80 degradation was observed at cold temperature over the time course of the study. The implementation of in-process hold time limits and temperature control in DS manufacturing has successfully resulted in an on-target PS80 concentration value in formulated mAb DS. Ultimately, the DS PS80 degradation mitigation and control strategies outlined in this work can significantly reduce the risk of particulates formation, and therefore, ensure product stability and consistency across all manufacturing batches.

207c Eliminating Batch-to-Batch Variability in Monoclonal Antibody Production Using Closed-Loop Control

Thomas Kavanagh, Maria Papathanasiou, Cleo Kontoravdi

Fed-batch bioreactors are the industry standard for the production of monoclonal antibodies (mAb) which are used for treatments of cancer. Most commercially available mAbs are produced by a mammalian cell-based system, namely Chinese Hamster Ovary (CHO) cells. Unlike bacterial cell cultures, mammalian cell-based systems allow for complex post-translational modification of the mAb to take place and lead to a more effective product. However, the complexity of these mechanisms and overall producing a product from a living organism leads to processes which exhibit increased batch-to-batch variability and inconsistent performance for example in productivity.

Current means of attaining a desired performance include open-loop control via optimisation of substrate feeding profiles (e.g., [1]). Unlike open-loop control, closed-loop control considers process feedback therefore compensation can be applied to account for process disturbances. Yet, there are very few examples of closed loop controllers for mammalian cell cultures. This is in part because of the highly non-linear nature of the process. In addition, process data is particularly expensive and laborious to generate. As such, model and control development remain challenging. Recent research in modelling and control of bacterial cell cultures has highlighted the advantages of closed-loop control (e.g., [2]). Similar advances are yet to be applied to mammalian cell-based systems.

In this work, an advanced control strategy is developed to track product titre by manipulating the feed of key metabolites to the cell culture. Using limited experimental data, a reduced-order model is developed for a controller describing only the manipulated, controlled, and constrained variables. From this, a model predictive controller is developed and the results from the simulation studies used to showcase the controller's performance. Future implementation is expected to reduce variability between batches.

[1] - Kotidis, Pavlos, et al. 'Model-based optimization of antibody galactosylation in CHO cell culture.' Biotechnology and bioengineering 116.7 (2019): 1612-1626.

[2] - Oh, Tae Hoon, et al. 'Integration of Reinforcement Learning and Model Predictive Control to Optimize Semi-batch Bioreactor.' AIChE Journal (2022): e17658.

207d Systematic Control Strategy Development for a Continuous Direct Compaction Line Via the Control Strategy Evaluation Tool (CET)

Jakob Rehrl, Julia Kruisz, Selma Celikovic, Peter Toson, Dalibor Jajcevic, Johan Remmelgas, Thomas O'Connor, Abdollah Koolivand, Geng Tian, Scott M. Krull, Johannes G. Khinast

Being able to track materials through a continuous manufacturing process is crucial to the implementation of a quality control strategy for solid oral dosage forms. Typically, residence time distribution (RTD) models are developed and parameterized for that purpose. RTD models allow for the prediction of the concentration of the active pharmaceutical ingredient (API) at specific locations in the manufacturing line. If the deviation in the API concentration from the setpoint exceeds a threshold, the generated material is considered out-of-specification (OOS) and needs to be discarded. The discharge of OOS material can then be triggered based on the output signal from the RTD model.

Any change in material attributes (e.g., due to a change of raw material batch) or process settings (e.g., due to an adjustment of the mass flow rate through the manufacturing line) can also change the RTD and affect the discharge decision. Therefore, a comprehensive knowledge of the critical factors and their impact on the RTD is essential during control strategy development. Because an extensive experimental characterization of the RTD is often not feasible, a simulation based on the discrete-element-method (DEM) is adopted. The RTD obtained from the DEM simulations at selected process settings or for specified material attributes is then approximated by suitable reduced order models and interpolation techniques, which significantly reduce the computational cost needed for a real time control strategy.

Finally, a control strategy evaluation tool (CET) is introduced, which has the functionality to test material discharge strategies in a systematic manner without running costly experiments. Three main components of the CET include: i) a modular simulation model of pharmaceutical manufacturing lines based on the RTD models mentioned above, ii) a library of typical disturbance scenarios, and iii) a discharge algorithm that implements a certain control strategy in terms of material discharge. The functionality of the CET is demonstrated using selected scenarios.

Disclaimer: This abstract reflects the views of the authors and should not be constructed to represent FDA's views or policies.

207e Modeling, Simulation and Control of Twin-Screw-Granulation and Fluid-Bed-Drying Applied to a ConsiGma^{TM}-25 Line

Selma Celikovic, Jakob Rehrl, Johannes Poms, Martin Horn, Johannes G. Khinast

Continuous pharmaceutical manufacturing offers several advantages compared to traditional batch manufacturing, such as increased production speed and efficiency. The ConsiGma^{TM}-25 is a continuous production line integrating a twin-screw-granulator (TSG), a fluid-bed-dryer (FBD), and a tablet press. To optimize process performance in terms of producing intermediates and products with precisely defined properties and to make the ConsiGma^{TM}-25 robust against process disturbances, adequate control concepts have been developed. This talk systematically outlines all the steps taken for the creation of the control concepts and it demonstrates their effectiveness by means of trial runs executed on a ConsiGma^{TM}-25 manufacturing line.

An essential component for the development of such control concepts is the real-time monitoring of critical quality attributes (CQA). The granule size after the TSG is considered to be an intermediate CQA. Therefore, a Parsum probe, capturing the particle size distribution (PSD) of wet granules, is placed at the TSG outlet. From this distribution data, relevant granule size characteristics are computed. A process model describing the relation between the granulation parameters and these characteristics has been developed by means of a local-linear model tree (LoLiMoT) algorithm. Model predictive control (MPC) uses this model as a core component and selects optimal granulation parameters in order to produce granules of well-defined PSD. The concentration of the active pharmaceutical ingredient (API) is another important CQA. This quantity is acquired by means of a Raman probe mounted after the TSG in combination with a model that computes the API concentration from the Raman spectral data. This information is fed to a feedback control concept that adjusts the process parameters in order to keep the API concentration close to its reference value. In parallel, the developed PSD model is utilized as a soft-sensor, allowing the real-time granule size monitoring, and acting as a potential replacement for the Parsum probe. Granule moisture after the FBD is a further CQA that should be monitored in real-time. For that purpose, a mathematical model based on physical laws has been developed and parametrized on the experimental data. This model runs in parallel with the process, estimating the granule moisture from the collected process data. Based on this estimation, a control concept stops the drying as soon as the desired moisture value has been reached. The performance of introduced control concepts has been examined on the ConsiGma^{TM}-25 plant by means of selected disturbance scenarios.

207g A Knowledge-Guided Framework for the Effective Application of Machine Learning Models in the Development of Condition Monitoring Systems.

Rexonni Lagare, Yan-Shu Huang, Zoltan Nagy, Gintaras Reklaitis

Machine learning models have suitable applications in the development of condition monitoring systems that ensure the quality, safety, and overall performance of a process. This preference stems from the black-box nature of machine learning models, where it is not necessary to have a well-established relationship between the predictions and the input variables, and all that is required are sufficient amounts of data. Often, this approach to developing condition monitoring systems can yield poor performances in detecting and diagnosing faults, and strategies for improvement are often unclear. Furthermore, the exclusion of process knowledge from the development process is counterproductive and could lead to misleading predictions.

In this work, we introduce a framework for the development of condition monitoring systems, where machine learning models are utilized based on knowledge of the process. This framework was applied to a pharmaceutical dry granulation line and was shown to be more effective than the traditional approach. Moreover, these advantages were quantified via metrics commonly used in machine learning, as well as proposed indices to quantify performance such as the ability of a condition monitoring system to classify novel faults.

411a Plug-and-Play Software for Mechanistic Modelling of End-to-End Continuous Manufacturing of Monoclonal Antibodies

Moo Sun Hong, Amos E. Lu, Richard D. Braatz

Monoclonal antibodies (mAbs) are the highest selling class of biopharmaceuticals. With continued growth in sales of existing mAb products and a growing pipeline of mAb product candidates being developed, the total sales of mAb products and all biopharmaceuticals will continue to increase in the coming years [1].

Biomanufacturing unit operations are largely designed and controlled through the use of data-driven models fit to experimental data. It is well-established that the use of mechanistic models can accelerate process development, simplify scaleup, and optimize manufacturing operations, but the available software tools have been limited. Mechanistic models for continuous biomanufacturing have been of increasing interest in recent years, as the industry strives to increase productivity and lower costs. Integrated mechanistic models for a continuous biomanufacturing plant need to account for the effects of changes in any unit operation on processes further downstream.

This presentation describes a software tool for carrying out high-fidelity dynamic simulations of integrated continuous mAb manufacturing plants. The simulations include mechanistic models of individual unit operations, including bioreactors, chromatography columns, viral inactivation units, and filtration units. The software implements the best mechanistic models available in the literature which are validated using experimental data. These validated individual units are then combined into a plant-wide dynamic model that includes the effects of model parameter uncertainties and disturbances, which is then used to validate the integration of the individual models and to map changes in operating conditions to the critical quality attributes and other variables of interest anywhere in the system.

In addition to design of the integrate continuous manufacturing plant, plant-wide mechanistic predictive models can be used to design, compare, and evaluate various control and real-time release testing strategies [2]. The plant-wide simulation software is designed to make the replacement of individual models and unit operations seamless. This plug-and-play ability enables the simulation-based evaluation of multiple process options before equipment is swapped in and out of the real physical manufacturing plant, including for less-established technology such as protein capture and purification via crystallization [3,4].

[1] M.S. Hong, W. Sun, A.E. Lu, and R.D. Braatz. Process analytical technology and digital biomanufacturing of monoclonal antibodies. *Am. Pharm. Rev.*, 23(6):122–125, 2020.

[2] M. Jiang, K.A. Severson, J.C. Love, H. Madden, P. Swann, L. Zang, and R.D. Braatz. Opportunities and challenges of real-time release testing in biopharmaceutical manufacturing. *Biotechnol. Bioeng.*, 114(11):2445–2456, 2017.

[3] M.S. Hong, K.A. Severson, M. Jiang, A.E. Lu, J.C. Love, and R.D. Braatz. Challenges and opportunities in biopharmaceutical manufacturing control. *Comput. Chem. Eng.*, 110:106–114, 2018.

[4] B. Smejkal, N.J. Agrawal, B. Helk, H. Schulz, M. Giffard, M. Mechelke, F. Ortner, P. Heckmeier, B.L. Trout, and D. Hekmat. Fast and scalable purification of a therapeutic full length antibody based on process crystallization. *Biotechnol. Bioeng.*, 110(9):2452–2461, 2013.

411b An Optimization-Based Approach to Identify Thermodynamically Stable Blends for Spray Drying Dispersions

Suela Jonuzaj, Malak Wehbe, Christopher Burcham, Amparo Galindo, George Jackson, Claire Adjiman

Many new compounds in the drug discovery pipeline are classified as poorly soluble chemicals which result in low bioavailability and dissolution rates. Spray drying dispersion (SDD) is a commonly used method in drug manufacturing to improve solubility, bioavailability and drug delivery of poorly soluble compounds [1]. SDD formulations consist of an active pharmaceutical ingredient (API), preferably in its amorphous state, dispersed and stabilized in a polymeric matrix to increase solubility. Despite the benefits of this technique, solid dispersions are often thermodynamically metastable. Therefore, it is important to select suitable API-polymer formulations that meet desired properties and ensure high solubility and long-term stability of the drug [2-4]. In current practice, heuristic approaches and experimental investigations are extensively employed for pre-screening small sets of commonly used polymers and solvents for the spray drying of APIs. Such protocols lead to high material consumption and thus they result in increased cost and environmental footprint. Clearly, there is an important need for developing more sophisticated and streamlined methods to reduce the number of time-consuming and costly experiments.

In this work, we present a systematic approach for identifying optimal (i) API-polymer and (ii) API-polymer-solvent formulations that meet desired physicochemical properties and can lead to drug products with improved bioavailability [5]. In particular, we employ property-prediction models to estimate solubility, miscibility and the glass transition temperature of a wide range of binary and ternary blends. In addition, we use optimisation models [6] to identify improved formulations that yield high solubility and stability of formulations with high drug loading. Multi-objective optimization is also utilized to model and identify optimal solutions of competing objectives. The design methodology is demonstrated with the help of a model drug, naproxen, where optimal polymers and solvents are selected to maximize drug loading while ensuring phase stability and a sufficiently high glass transition temperature of the final blend. A ranked list of optimal blends with different chemicals and compositions is obtained by introducing integer cut inequalities into the model, and proper phase diagrams of the best binary and ternary mixtures are constructed. Finally, the proposed model is used to generate a library of optimal formulations for various SDD systems that is used to identify better-performing designs and guide experimental work.

[1] Duarte, I., Santos, J.L., Pinto, J.F., Temtem, M., 2015, Screening methodologies for the development of spray-dried amorphous solid dispersions, Pharmac. Research 32, 222-237.

[2] Newman, A., Zografi, G., 2022, What Are the Important Factors That Influence API Crystallization in Miscible Amorphous API−Excipient Mixtures during Long-Term Storage in the Glassy State?, Molecular Pharmaceutics 19, 378-391.

[3] Lehmkemper, K., Kyeremateng, S.O., Heinzerling, O., Degenhardt, M., Sadowski, G., 2017, Impact of polymer type and relative humidity on the long-term physical stability of amorphous solid dispersions, Molecular Pharmaceutics 14, 4374-4386.

[4] Dohrn, S., Reimer, P., Luebbert, C., Lehmkemper, K., Kyeremateng, S.O., Degenhardt, M., Sadowski, G., 2020, Thermodynamic modeling of solvent-impact on phase separation in amorphous solid dispersions during drying, Molecular Pharmaceutics 17, 2721-2733.

[5] Jonuzaj, S., Burcham, C.L., Galindo, A., Jackson, G., Adjiman, C.S., 2022, Optimizing the selection of drug-polymer-solvent formulations for spray-dried solid dispersions in pharmaceutical manufacturing, Computer Aided Chemical Engineering, Accepted for publication.

[6] Jonuzaj, S., Cui, J., Adjiman, C.S., 2019, Computer-aided design of optimal environmentally benign solvent-based adhesive products, Computers andamp; Chemical Engineering 130, 106518.

411c Impact of Excipient Properties and Blending Process Parameter on Blend Uniformity of a Blend Prepared in Small Batches Semicontinuously Using Integrated Feeder-Blender System

Sumit Kumar, Gintaras V. Reklaitis, Zoltan Nagy, Marcial Gonzalez, Paul R. Mort

In the pharmaceutical industry, solid dosages have multiple powder ingredients such as active pharmaceutical ingredients (API), excipients, lubricants etc. Mostly all the oral solid manufacturing processes start with the feeding of these ingredients followed by blending unit operation. The initial powder feeding step defines the amount of a specific ingredient entering and thus determines the final content of a formulation. Hence it is important to add the right quantities in the right proportions and to have the required blend homogeneity in powder to ensure accurate dosing of API in the final dose[1]. Inconsistent feeding and blending can lead to non-conforming and out of specification products and may lead to batch failure.

The manufacturing process can be in a batch, continuous or semi-continuous mode of manufacturing Previous work talks about the comparison of batch and continuous blending and the benefits and issues associated with both[23]. Another approach can be to prepare a series of small batches. Output from these small batches can be delivered downstream to both batch or continuous modes and have the flexibility of customizing the batch sizes. It can feed the blended material to downstream semicontinuous unit operations such as fluidized bed granulators and dryers. Such an approach has the added benefit of having relatively easier scaleup and flexibility in batch size as it can be varied by changing the number of small batches. The feeding-blending operation is the major source of variance in the final product quality in continuous manufacturing[4]. The pharmaceutical industry is constantly innovating[5] and researchers are looking for improved, robust and effective integrated feeding blending solutions.

The present study aims to use an integrated feeder blender system by Syntegon to provide output as a series of small batches semi-continuously. The Xelum Randamp;D feeder blender module from Syntegon is operated by one to five gravimetric feeders for dosing different powder materials into the conical blender. The blender then mixes the ingredients as per the set recipe and operating parameters. The output of the integrated feeder blender system is then unloaded in small batches. The blend uniformity of the unloaded material is checked using a UV visible spectrophotometer[6]. Batch to batch variation and within batch variations using such a system will then be analyzed. Blender operating parameters such as blending time, stirrer rotation speed (rotation per minute (RPM)) and fill level will be varied based on a DOE design. The research will also investigate the impact of material properties of the ingredients such as particle shape, particle size, density and flow properties on the blend uniformity of the API in the powder mixture. This work showcases the combining of the two most common unit operations and providing the output that could be used for both batch and continuous modes. Such an integrated process can eliminate the use of manual feeding and long batch blend times. Also, this setup provides the option to have a wide range of residence time that may be needed to blend very few to many ingredients in one step. The current study investigates the performance of blending of the integrated feeder blender system and its usage as an alternative to batch blending in rotary blenders and continuous blending in tubular blenders.

Reference:

1. Cullen PJ, Romañach RJ, Abatzoglou N, Rielly CD. *Pharmaceutical Blending and Mixing.*; 2012. doi:10.1002/9781118682692
2. Jaspers M, de Wit MTW, Kulkarni SS, et al. Impact of excipients on batch and continuous powder blending. *Powder Technol.* 2021;384:195-199. doi:10.1016/j.powtec.2021.02.014
3. Alyami H, Dahmash E, Bowen J, Mohammed AR. An investigation into the effects of excipient particle size, blending techniques andamp; processing parameters on the homogeneity andamp; content uniformity of a blend containing low-dose model drug. *PLoS One*. 2017;12(6):1-19. doi:10.1371/journal.pone.0178772

4. Zoltan K. Nagy, Arwa. El Hagrasy JL. *Continuous Pharmaceutical Processing*. 1st Editio. Cham : Springer International Publishing : Imprint: Springer; 2020.
5. Rogers A, Ierapetritou M. Challenges and opportunities in modeling pharmaceutical manufacturing processes. *Comput Chem Eng*. 2015;81:32-39. doi:10.1016/j.compchemeng.2015.03.018
6. Al-obaidi ZMJ, Ebrahim AA, Ani A, Al-juhaishi AMR, Hussein AN. A comparative study for the quantification of paracetamol in multicomponent oral solution employing standard addition method utilized in UV-Visible spectroscopy and RP-HPLC. *J Pharm Sci Res*. 2019;11(2):339-342.

coperion
coperion
METTLER TOLEDO
SYNTEGON
Xelum R&D

411d Molecular Dynamics Modeling Based Investigation of the Effect of Freezing Rate on Protein Stability

Tibo Duran, Bruna Minatovicz, Ryan Bellucci, Jun Bai, Bodhisattwa Chaudhuri

Purpose:

Bulk protein therapeutics are generally frozen for long term storage to avoid product degradation and facilitate product transportation. As a result, freezing of water induces solute rejection, creating regions of high solute concentration, which may impact protein to be destabilized and lose its activity. In addition, the stability of protein drug products frozen during fill finish operations is greatly affected by the freezing rate applied. Non-optimal freezing rates may lead to the denaturation of protein`s complex macromolecular conformation. However, limited work has been done to address the effect of different freezing rates on protein stability at nano-scale level. The stability of a model protein, lysozyme, was investigated at atomic and molecular scale under varying freezing rates and moving ice-water interface using Molecular dynamics (MD) simulations. Ice seeding approach was adopted to initiate ice formation in this present simulation. The simulation results from this current work were further validated with previous experimental data on freezing bulk protein therapeutics.

Methods:

Lysozyme protein solution in 10 mM phosphate buffer at pH 7.0 was chosen as initial conditions applied into this current MD simulations. The MD simulation trajectories and their analyses were carried out using GROMACS, VMD and Pymol software. All-atom MD simulations were applied using CHARMM36 force field with temperatures gradually decreased from -3˚C to -26 ˚C with 5 – 6 ˚C temperature intervals. The initial protein crystal structure of LYZ was taken from the protein data bank entries (pdb: 1AKI). The protein was initially modeled using the psfgen tool in VMD, and then further solvated in a rectangular simulation box with the charge neutralized with $NaPO_4$. According to experimental work, an ice seeding approach was adopted to initiate ice formation in the MD simulation work. The simulation system was equilibrated in the NPT ensemble by initially applying 1000 cycles of a conjugate-gradient minimization scheme followed by a short 500-ps MD run in NPT assemble. The temperature was controlled using the Langevin thermostat, and the pressure was controlled by the Nose-Hoover barostat. The simulations were carried out using periodic boundary conditions. This briefly equilibrated system of lysozyme protein buffer solution was further subjected to MD simulation sets in the NVT ensemble. Computations were performed on HPC, a supercomputer cluster at the University of Connecticut.

Results:

Ice formation was successfully initiated by ice seeding approach in the model protein buffer solution under freezing process. The faster freezing rate (11-12 K/490 ns) applied resulted in overall smaller ice fraction within the simulation box with a larger freeze-concentrated liquid (FCL) region. Consequently, the faster freezing rate better maintained protein stability with less secondary structure deviations, higher hydration level and structural compactness, and less fluctuations at individual residues than observed following slow (5-6 K/490 ns) and medium (7-8 K/490 ns) freezing rates. The present study also identified the residues near and within lysozyme protein helices 3, 6, 7, and 8 dominate the structural instability of the lysozyme at 247 K freezing temperature. Moreover, after the freezing simulations, the value of lysozyme concentration increased 5.39-fold compared with the concentration at the beginning of the simulation, which is comparable with the experimental result (5.41-fold increase).

Conclusions:

The good agreement between the computational predictions and experimental results highlights the benefit of the all-atom MD approach to gain insight on ice formation and protein behavior during freezing. To the best of our knowledge, different freezing rates were studied "non-isothermally" for the first time in this current work. Thus, a good understanding of freezing rates on protein instability was revealed by applying the developed computational model, which can be utilized as an effective and powerful tool in the process of discovery, development, and optimization of new protein therapeutics.

Keywords:

freezing rate, ice-water interface, protein stability, molecular dynamics simulations

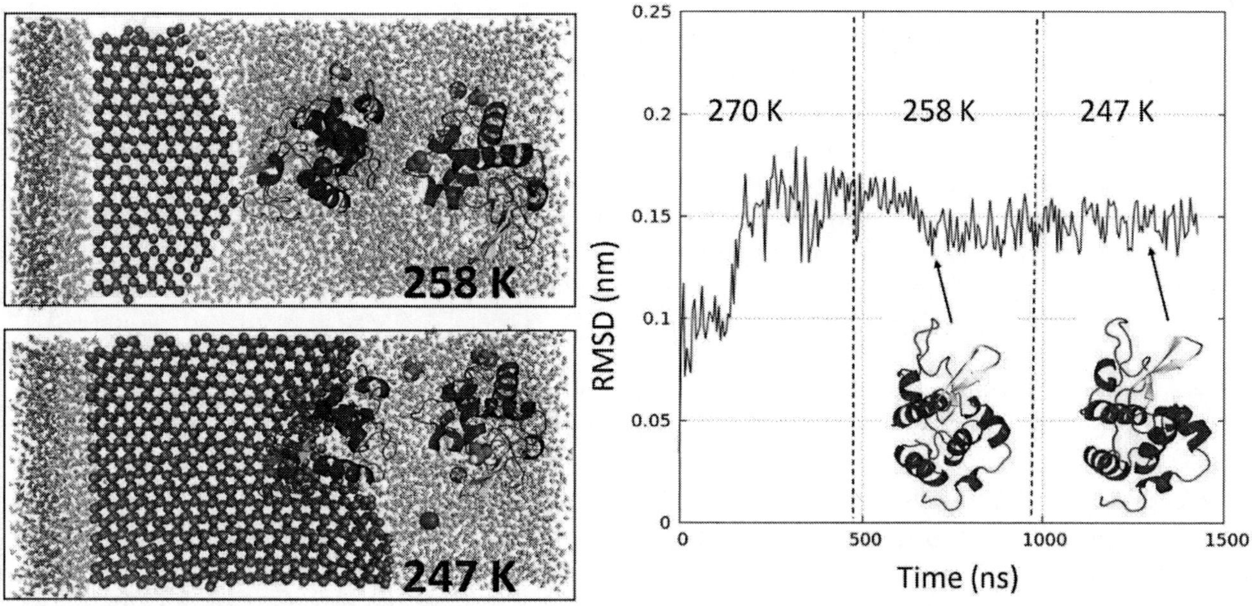

411e A Multi-Scale Approach to Understanding Spray Dried Dispersion Impurity Generation

James Miesle, Salvador Salvador Garcia, Matthew Walworth, Jason Melnick, Lars Magnusson, Alana Kurz, Kerri Fischer

A common strategy to enhance the bioavailability of poorly soluble molecules is to generate an amorphous solid dispersion by spray-drying. This processes generally requires the dissolution of the drug molecule in an organic solvent, which is then spray-dried co-currently by means of a hot nitrogen gas stream to remove the majority of the solvent phase. Typical low boiling point organic solvents, such as acetone, dichloromethane, ethanol, or methanol, are often used due to the suitable solubility of drug substance. Larger scale spray dryers used to support late phase clinical and commercial production typically operate in a closed loop configuration. Residual amounts of gas phase solvent are recirculated and exposed to elevated temperatures during reheating of the gas stream to the desired process conditions.

Amine functional groups, often found in new drug molecules, are known to undergo nucleophilic substitution reactions promoted by aldehydes or ketones. Primary alcohols such as ethanol or methanol can undergo dehydrogenation in the presence of a metal oxide catalyst at high temperature and low pressure, yielding the corresponding aldehyde. Data suggests that these conditions exist in the heater of the recirculated drying gas in a spray drying process; sufficiently to generate measurable quantities of the aldehyde. The resulting aldehyde can react further to promote the chemical degradation of the amine group in the drug molecule. Furthermore, the presence of the aldehyde in some systems is known to induce strong non-linearities in the vapor-liquid equilibria by the generation of new species resulting from the interaction of the solvent and the generated aldehyde. This work studies the extent of formaldehyde formation and the speciation consequences in a closed-loop spray drying process in the presence of a stainless-steel heater, while spray-drying a methanol solution.

411f Parallel Process Development and Enzyme Evolution for Drug Substance Manufacturing

Sean Dubina, Scott McCann, Birgit Kosjek, Stephanie Chun, Nilusha Padivitage, Daniel DiRocco, Jonathan McMullen

Process development of a Dynamic Kinetic Resolution (DKR) step in the commercial synthesis of an API intermediate was undertaken in preparation for large-scale batch processing. By utilizing a Ketoreductase (KRED) enzyme in lieu of a Ru catalyst to catalyze the DKR reaction, the process could be streamlined to utilize a greener, safer, and more robust chemo-enzymatic through-process over the Ru-catalyzed route.

To support successful scale-up of this process, process development studies with cross-collaborative teams were performed at bench, kilo, and pilot plant scales in parallel with KRED evolution. The initial KRED variant limited process parameters to a narrow, stringent process. By understanding the process sensitivities of the enzyme such as organic solvent tolerance, pH, and temperature, a robust process was optimized that provided the necessary conversion of starting material while minimizing risk of enzyme degradation and generation of impurities. The DKR reaction was scaled up in the pilot plant at Merck andamp; Co., Inc., Rahway, NJ, USA to understand scale-up behavior and risks. Key takeaways from the first batch provided valuable insight into important aspects of the process that required further investigation and improvement. The cross-functional team worked closely with internal and external partners to demonstrate a robust process that could be implemented in the pilot plant and eventually, the manufacturing production plant. Evolving the enzyme while developing a streamlined DKR through-process using common equipment was paramount to successful tech transfer of this reaction across sites and eventually, process characterization.

411g An Industrial in silico Modeling Platform for Reliable, Efficient, and Agile Process and Product Development

Fabrice Schlegel, Pablo A. Rolandi, Iman Mirzaee, Maxwell Maritato, Saman Seifi, Miguel Angel Valderrama-Gomez, Joao De Faria, Saeed Jafari Kang, Maryam Medghalchi, Elcin Icten Gencer, Justin Porth

Platforms are a force for innovation in Process Development at Amgen and In Silico Modeling is a new digital platform that complements established product and process platforms. Ad hoc modeling and models is inefficient, error prone, does not allow for enterprise ready solutions, and its insight is often not credible in a regulatory context. Platform thinking is essential to digital innovation and our journey towards digital twins.

This talk will discuss the current state of the Amgen's In Silico modeling platform and its four main pillars:

1. Modeling assets: the collection of in silico models and data deployed to and used by engineers and scientists across the organization.
2. Modeling technologies: the systems and tools that underpin the above.
3. Modeling processes: the collection of business processes that define the lifecycle of the platform and its components.
4. Modeling maturity: a capability maturity framework for enterprise grade modeling initiatives

This work will share industrially relevant lessons on how to scale-up the impact of in silico modeling teams. It will also present practical Agile modeling practices to enable agility and speed. Finally, it will also discuss the role of strategic partnerships within the organization and with the wider ecosystem of vendors, academia, and stakeholders.

567a Pharma 4.0: Advanced Continuous Pharmaceutical Tablet Manufacturing Process

Ravendra Singh

Currently, Industry 4.0 concepts are being applied to pharma industry to achieve Pharma 4.0 paradigm. Pharma 4.0 reduces the time and resources needed for continuous pharmaceutical manufacturing and also improves the product quality and production consistency. It has many advantages but also have bigger challenges on the applications of artificial intelligence (AI)/machine learning (ML), material traceability, optimization, advanced process control, and cyber-physical security side because of the different levels of complexities involved [1-3]. The predictive capabilities and the quality of the pharmaceutical products can be improved significantly via employing the artificial intelligence and the advanced model predictive control (MPC) system if an appropriate cyber-physical security defense is in place.

In this work, five components of industry 4.0 namely artificial intelligence (AI)/machine learning (ML), material traceability, optimization, advanced control, and cyber-physical security, have been developed and implemented into the continuous pharmaceutical manufacturing process.

Four machine learning (ML) models have been trained to predict the response of continuous pharmaceutical manufacturing process and the performance of these ML models has been compared. The investigated ML methods are long short-term memory (LSTM), 1D convolution neural network (CNN), random forest (RF), and artificial neural network (ANN). The best performing ML model is then implemented into the continuous pharmaceutical tablet manufacturing process for real time prediction.

A systematic framework including the methods and tools have been developed for material traceability. A corresponding software tool has been also developed to automate the material traceability procedure. The heart of the material traceability toolbox is the RTD model. The applications of the developed methods and tools have been demonstrated for material traceability of continuous pharmaceutical manufacturing process.

A systematic framework including the methods and tools has been developed for dynamic optimization of the feeder refill strategies. The deviation of the outlet mass flow of the feeder from the targeted flow rate has been minimized to obtain the optimum value of the feeder refill parameters. The material properties also affect the refill strategy meaning that the feeder refill strategy need to be frequently optimized if there are any changes in the materials and plant. Therefore, the developed feeder model and dynamic optimization tool can save the time and recourses as well as can improve the product quality significantly.

An advanced model predictive control (MPC) system has been implemented in the continuous pharmaceutical manufacturing (CPM) pilot-plant. The CPP's and CQA's are controlled in real time. The critical control variables that have been controlled using model predictive control (MPC) system are drug concertation, powder level before tablet press, main and pre compression forces, tablet weight and hardness. A novel control strategy for powder level control in a chute placed in between blender and tablet press unit operation of continuous tablet manufacturing process has been developed, implemented and evaluated.

A systematic framework including the methods and tools have been also developed for proactive identification and mitigation of potential cyber-physical attack risk on continuous pharmaceutical manufacturing plant [2]. The cyber-physical security relevant software tools such as Snap 7, Wireshark, and Tripwire have been applied to CPM. A novel software tool named CPS (Cyber-Physical Security) has been developed for cyber-physical security of the continuous pharmaceutical manufacturing. The integrated commercially available and developed (in house) cyber-physical security tools have added an extra layer of security of our continuous pharmaceutical manufacturing pilot-plant for any unexpected attacks.

All the relevant data generated during continuous manufacturing has been systematically collected, stored and organized in a data hub (OSI PI) and cloud system as per industry 4.0 standard.

References

1. Bhaskar, A., Barros, F. N., Singh, R. (2017). Development and implementation of an advanced model predictive control system into continuous pharmaceutical tablet compaction process. International Journal of Pharmaceutics, 534 (1-2), 159-178.
2. Billups, M., Singh, R. (2018). Systematic framework for implementation of material traceability into continuous pharmaceutical tablet manufacturing process. Journal of Pharmaceutical Innovation. https://doi.org/10.1007/s12247-018-9362-9.
3. Singh, R. (2021). Cyber-Physical Security (CPS) Tool for Continuous Pharmaceutical Manufacturing Process. Pharma Focus Asia. Issue 44, 41-49. https://www.pharmafocusasia.com/information-technology/cyber-physical-security-tool.

567b Development of a DEM Based Digital Twin of a Tablet Press Feed Frame for the Evaluation of Continuous Manufacturing Control Strategies.

Johan Remmelgas, Dalibor Jajcevic, Peter Toson, Marko Matic, Michela Beretta, Jakob Rehrl, Julia Kruisz, Thomas O'Connor, Abdollah Koolivand, Geng Tian, Scott M. Krull, Johannes G. Khinast

Pharmaceutical process development can require a substantial amount of material for experimental characterization, and experimental work is therefore typically restricted to a small design space (e.g. narrow material variation and operating range). In order for the industry to explore a larger design space and carry out more optimization studies, *in silico* modelling and high-fidelity digital twins offer the possibility to perform numerous virtual experiments.

The Research Center Pharmaceutical Engineering (Austria), supported by the U.S. FDA, have recently started a project to develop an approach for assessing drug manufacturing control strategies based on high-fidelity digital twins, taking a direct compression line as a use-case. In this high-fidelity digital twin platform, typical pharmaceutical materials, equipment, and processes were simulated using verified numerical methods and algorithms [1,2,3].

This work presents simulations of a tablet press feed frame using the discrete-element-method (DEM) model. Due to the large hold-up masses and mean residence times in the feed frame, it was computationally unreasonable to resolve the full Residence Time Distribution (RTD) with DEM simulations. Therefore, a DEM extrapolation technique colloquially called 'Relay Race' was used to predict the full RTD. The RTD obtained in this fashion was then used in a reduced order model to predict responses to various disturbances at different feed frame throughputs and tablet press speeds. These predictions were compared to experimental data. Analysis of the DEM data provided a deeper insight into the process, revealing flow patterns that are very challenging to determine experimentally.

Disclaimer: This abstract reflects the views of the authors and should not be constructed to represent FDA's views or policies.

References

[1] Siegmann *et al.*, 2020. *Powder flow and mixing in different tablet press feed frames*. Adv Pow Tech 31, 770–781. https://doi.org/10.1016/j.apt.2019.11.031

[2] Toson *et al.*, 2019. *Explicit Residence Time Distribution of a Generalised Cascade of Continuous Stirred Tank Reactors for a Description of Short Recirculation Time (Bypassing)*. Processes 7, 615. https://doi.org/10.3390/pr7090615

[3] Siegmann *et al.*,2021. *Massively speeding up DEM simulations of continuous processes using DEM extrapolation*. Pow Tech 390, 442–455. https://doi.org/10.1016/j.powtec.2021.05.067

567c Integrated System Modeling That Connects the Drug Product Manufacturing Process to Dissolution Testing

Indu Muthancheri, Marilyn Calderone, Nicolas Sanchez, Cecile Gabaude-Renou, Robert Yule

The quality-by-design paradigm in the pharmaceutical industry has been systematically incorporating science-based process understanding in all manufacturing stages. This increased interest in integrating mechanistic process models into the development workflow has led to improved product quality, reduced consumption of raw materials, reduced time to market, and reduced risk [1,2]. Adapting smart manufacturing also referred to as Industry 4.0 practices, provides further digitalization opportunities involving the application of model-based methodologies to the pharmaceutical industry [3].

Process system modeling or flowsheet modeling evaluates mechanistic models for unit operations when connected in series. Considering that a manufacturing process consists of multiple unit operations in series, a system modeling approach can aid design space exploration and optimized product quality testing.

In drug product manufacturing, the dry granulation process is commonly used for moisture-sensitive and highly cohesive powders. The manufacturing route consists of multiple unit operations: feeder, roller compactor, mill, and tablet press. The importance of a process system model in providing a fundamental understanding of the dry granulation process with limited experimental calibration has been demonstrated previously by Gavi, Emmanuela, and Gavin K. Reynolds [4].

In the presented work, the impact of material property in the dry granulation process system model is studied. The following improvements are made to the existing workflow [4].

i. A material sparing approach developed by Sousa, Ricardo, et al. [5] is used to model the roller compaction process. This is a simplified Johanson model where the ribbon strength is independent of the nip angle, thereby removing the need to measure the effective angle of internal friction and angle of wall friction of the feed material.
ii. The impact of ribbon porosity on the screen mill model parameters is evaluated to predict the particle size distribution. The results show that the fines fraction in the bimodal fragment distribution function is highly sensitive to the ribbon porosity.
iii. A bi-modal particle size distribution is also used to predict the tablet disintegration and dissolution process. The study compares the deviation of model prediction when considering a uniform, unimodal and bimodal particle generation during the tablet disintegration process.

A detailed analysis of multiple unit operations, such as roller compactor, screen mill, tablet press, and QC dissolution, provides the effect of each process parameter on final tablet product performance. The calibrated ‘digital twin’ is further utilized in decision-making, including evaluation of design and scale-up decisions, process optimization at the pilot-scale, sensitivity analysis, and technology transfer.

[1] Hakemeyer, Christian, et al. 'Process characterization and design space definition.' Biologicals 44.5 (2016): 306-318.

[2] Rogers, Amanda, and Marianthi Ierapetritou. 'Challenges and opportunities in modeling pharmaceutical manufacturing processes.' Computers andamp; Chemical Engineering 81 (2015): 32-39.

[3] Litster, James, and Ian David L. Bogle. 'Smart process manufacturing for formulated products.' Engineering 5.6 (2019): 1003-1009.

[4] Gavi, Emmanuela, and Gavin K. Reynolds. 'System model of a tablet manufacturing process.' Computers andamp; chemical engineering 71 (2014): 130-140.

[5] Sousa, Ricardo, et al. 'Roller compaction scale-up made simple: an approximate analytical solution to johanson's rolling theory.' Journal of Pharmaceutical Sciences 109.8 (2020): 2536-2543.

[6] Process Systems Enterprise, gPROMS, www.psenterprise.com/products/gproms, 1997-2020

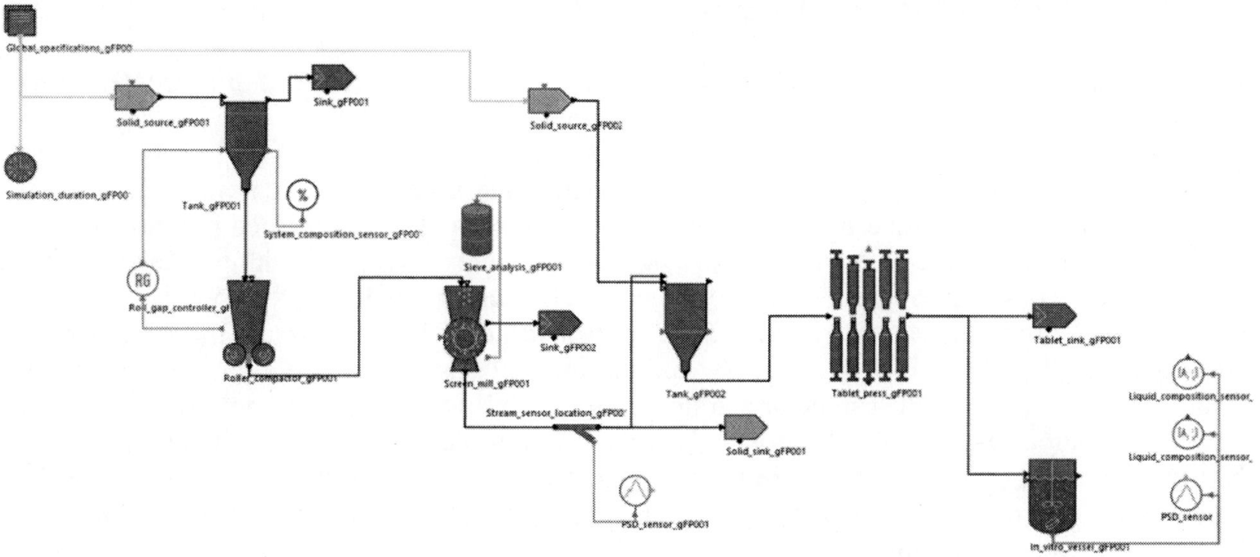

Figure: gPROMS [6] flowsheet model for drug product manufacturing using dry granulation process

567d Hybrid Machine Learning Assisted Model Predictive Control of a Continuous Dry Granulation Tableting Line

Yan-Shu Huang, Rexonni Lagare, Sunidhi Bachawala, Marcial Gonzalez, Zoltan Nagy, Gintaras Reklaitis

Active control strategies are playing an essential role in modern pharmaceutical manufacturing. Instead of controlling the process input variables within tight ranges as practiced in traditional control approaches, a more flexible active control strategy has the potential to deal with process uncertainties and variations in raw material properties. Model predictive control (MPC) often serves as the preferred candidate for a centralized plant-wide control strategy. When there are many constraints on critical quality attributes and strong process variable interactions exist in the multiple-input multiple-output (MIMO) system, MPC can demonstrate better capabilities in setpoint tracking and disturbance rejection compared to classical PID control. However, the implementation of MPC in pharmaceutical tablet manufacturing processes is still in infancy. Most work reported in the literature is limited to a single process unit or the direct compaction process; integrated dry granulation or wet granulation tableting processes still require further research [1].

Dry granulation via roller compaction (RC) is increasingly applied in the pharmaceutical industry to change particle size distribution (PSD) of in-process powders or final products. The granulation step is beneficial for improving powder flowability and addresses problems related to dustiness, low bulk density, and poor blend uniformity [2]. Nevertheless, unless the process is well-controlled, when granules are subsequently compressed into tablets, the powder compressibility may be compromised, and tablet tensile strength is sacrificed [3]. Therefore, a model to correlate the properties of powders, ribbons, granules, to those of the tablets produced is required to understand and further control the process appropriately. Mechanistic models such as Johanson's model [4] and Reynold's model [5] are typically used to describe roll compaction, while population balance models (PBM) can account for the milling step. However, it is complicated to determine the breakage function in the PBM purely based on ribbon fracture physics. Machine learning (ML) is a preferred data-driven alternative to developing a mechanistic model. Moreover, ML and mechanistic model components can be combined into a hybrid model to maintain high physical interpretability and feasibility.

This study is focused on developing and implementing model predictive control in continuous dry granulation tableting processes. First, a hybrid ML-assisted model is reported, which serves to predict the process behavior and the essential product attributes. Process behavior of the rotary tablet press and tablet properties are described by a mechanistic model which has been implemented in previous work [6]. Neural network and genetic programming approaches are used to predict the PSD of granules produced by the roller compactor, as well as to map the impact of granule PSD and roll compaction on tablet properties. In the second part, the hybrid ML-assisted MPC, implemented both in silico and in our physical continuous tableting pilot plant, are described. The results of investigating alternative control strategies and their real-time control performance will be summarized.

[1] Jelsch, M., Roggo, Y., Kleinebudde, P., andamp; Krumme, M. (2021). Model predictive control in pharmaceutical continuous manufacturing: A review from a user's perspective. *European Journal of Pharmaceutics and Biopharmaceutics*, *159*, 137-142.

[2] Toson, P., Lopes, D. G., Paus, R., Kumar, A., Geens, J., Stibale, S., ... andamp; Khinast, J. (2019). Model-based approach to the design of pharmaceutical roller-compaction processes. *International*

journal of pharmaceutics: X, 1, 100005.

[3] Herting, M. G., andamp; Kleinebudde, P. (2008). Studies on the reduction of tensile strength of tablets after roll compaction/dry granulation. *European journal of pharmaceutics and biopharmaceutics*, *70*(1), 372-379.

[4] Johanson, J. R. (1965). A rolling theory for granular solids.

[5] Reynolds, G., Ingale, R., Roberts, R., Kothari, S., andamp; Gururajan, B. (2010). Practical application of roller compaction process modeling. *Computers andamp; chemical engineering*, *34*(7), 1049-1057.

[6] Huang, Y. S., Sheriff, M. Z., Bachawala, S., Gonzalez, M., Nagy, Z. K., andamp; Reklaitis, G. V. (2021). Evaluation of a Combined MHE-NMPC Approach to Handle Plant-Model Mismatch in a Rotary Tablet Press. *Processes*, *9*(9), 1612.

567e Smart Process Analytics for the Prediction of Critical Quality Attributes in End-to-End Batch Manufacturing of Monoclonal Antibodies

Moo Sun Hong, Fabian Mohr, Chris Castro, Tom Mistretta, Roger A. Hart, Ben Smith, Richard D. Braatz

Biopharmaceuticals are continuously growing in terms of both global sales and their percentage of the overall pharmaceutical pipeline [1]. Especially, monoclonal antibodies are number two in the number of drugs in the current pipeline, which is after the general class of small molecules produced via traditional synthetic chemistry techniques [2].

Depending on the degree of process understanding, the process models range from data-driven to mechanistic [3]. For many processes for manufacturing of biopharmaceuticals, mechanistic models are not available. This situation has required manufacturers to develop data-driven models using data analytics (DA)/machine learning (ML) methods [4]. Past studies have already demonstrated that DA/ML methods can construct accurate and reliable models for industrial biomanufacturing processes by identifying and building relationships between critical process parameters (CPPs) and CQAs [5,6]. Since the available DA/ML tools and software packages are rapidly increasing today, the challenge is how to *select* the best methods and tools for a specific biomanufacturing dataset to ensure that the most accurate and reliable model is constructed. The optimal selection of DA/ML tools requires a substantial level of expertise due to diverse nature of biomanufacturing data in terms of quantity and quality.

This challenge motivated the development of a smart process analytics (SPA) software, which automates the robust selection of methods and construction of models [5,6,7]. With the given datasets, the software first assesses the specific data characteristics such as nonlinearity, multicollinearity, and dynamics. Based on the data characteristics and the user objective, the software selects the most suitable DA/ML tool by following a decision tree which can be represented in the form of a triangle. The SPA software constructs the final model using a rigorous cross-validation procedure for optimal tradeoff between robustness (variance) and accuracy (bias). This presentation describes learnings obtained by the application of smart process data analytics to the prediction of critical quality attributes from biomanufacturing datasets. The automated determination of the best DA/ML tools for model construction and process understanding in SPA is validated by comparison to alternative DA/ML methods – including partial least squares, ridge regression, elastic net, and sparse partial least squares – and we describe the results of a detailed residual analysis that identifies biases and other deviations from method assumptions that occur in applications to such datasets. Motivated by these residual analyses, we describe extensions to SPA that improve the predictive accuracy while accounting for variations that occur on longer time scales than considered in past models.

[1] M.S. Hong, K.A. Severson, M. Jiang, A.E. Lu, J.C. Love, and R.D. Braatz. Challenges and opportunities of biopharmaceutical manufacturing control. *Computers andamp; Chemical Engineering*, 110:106-114, 2018.

[2] I. Lloyd. *Pharma Randamp;D Annual Review 2021*. London, UK: Informa, 2021.

[3] Basic Principles of GMP. Geneva, Switzerland: World Health Organization, 2008.

[4] M.S. Hong, W. Sun, A.E. Lu, and R.D. Braatz. Process analytical technology and digital biomanufacturing of monoclonal antibodies. *American Pharmaceutical Review*, 23(6):122-125, 2020.

[5] K. Severson, J.G. VanAntwerp, V. Natarajan, C. Antoniou, J. Thömmes, and R.D. Braatz. A systematic approach to process data analytics in pharmaceutical manufacturing: The data analytics triangle and its application to the manufacturing of a monoclonal antibody. In: *Multivariate Analysis in the Pharmaceutical Industry*, London, UK: Academic Press, 295-312, 2018.

[6] K. Severson, J.G. VanAntwerp, V. Natarajan, C. Antoniou, J. Thömmes, and R.D. Braatz. Elastic net with Monte Carlo sampling for data-based modeling in biopharmaceutical manufacturing facilities. *Computers andamp; Chemical Engineering*, 80:30-36, 2015.

[7] W. Sun and R.D. Braatz. Smart process analytics for predictive modeling. *Computers andamp; Chemical Engineering*, 144:107134, 2021.

567f Characterizing the Freezing Process of Biopharmaceuticals in Vials at Commercial Scale: COVID-19 Vaccine Case Study and Mechanistic Modeling

Leif-Thore Deck, David R. Ochsenbein, Marco Mazzotti

Keywords: Freezing, Freeze-Drying, Lyophilization, Nucleation, Modeling, Quality by Design

The role of freezing in the pharmaceutical industry has recently garnered substantial public interest in the rise of the COVID-19 pandemic. Storage conditions of COVID-19 vaccines have received global news coverage, with low temperature cold chains extremely costly or unavailable depending on the market. Although the stabilizing effect of freezing has been studied extensively in the past, its quantification and prediction remain challenging until today.

The stability of a frozen drug product depends not only on its physicochemical properties, but also on the way it is frozen, i.e. on how it nucleates and solidifies. Ice nucleation, the first step of freezing, is a stochastic process whose effect is of particular importance for small volumes such as the ones used in filled vials. It induces variability in both nucleation and solidification properties among frozen products, with potential implications on stability and processability. Predicting this heterogeneity among frozen products therefore is of great interest for freezing process design and optimization. Unfortunately, no model available in literature is capable of such prediction, likely due to the complexities of the production scale setups used in industry: Typically, tens of thousands of vials are densely packed together in a pallet and frozen slowly in a cold storage room. Such configuration results in enhanced variability due to thermal interaction among the vials, as well as due to spatial differences in heat transfer across the pallet.

To tackle this challenge, we developed a mechanistic modeling framework that computes the freezing behavior of all vials in an arbitrarily sized batch. It takes into account the stochasticity of ice nucleation via a Monte Carlo approach. We compare the simulation findings with experimental data obtained from engineering runs of the Janssen COVID-19 vaccine, for which the model qualitatively reproduced all observed trends. To promote the use and further development of the model in the wider community, we provide open source access to it in the form of a python package (available at https://pypi.org/project/ethz-snow/).

Acknowledgement

The authors thank the Janssen Pharmaceutical Companies of Johnson andamp; Johnson for the support in the course of the project and the financial funding. We especially thank Juan Carlos Araque and Ryan Wall.

567g Smart Data Analytics for Fault Detection and Its Application to Biopharmaceutical Manufacturing

Fabian Mohr, Weike Sun, Richard D. Braatz

Maintaining a high quality of the product is a key factor in most manufacturing processes. In order to ensure quality and maintain safe operations, process monitoring schemes are employed. Fault detection, which is the detection of abnormal operating conditions, is the first step in a chain of steps in process monitoring systems that need to be performed [e.g., see Refs. 1 to 4 and citations therein]. Many new data analytic tools have been introduced in recent years for a variety of objectives including fault detection. The increasing multitude of these powerful tools has motivated the development of smart data analytics approaches, i.e., a decision tree that automatically selects the most suitable method based on metadata and a systematic interrogation of the dataset [5].

In contrast to a past journal article [5] that developed a smart process analytics approach for building data-driven models for process optimization and control, this presentation builds such a decision tree for the objective of process monitoring. More specifically, we develop an automatic method selection strategy for the objective of *fault detection*, to select and apply the most suitable fault detection algorithm for a given problem. This decision tree is constructed based on (1) theoretical analyses of methods for extracting information on the data characteristic and on the mathematical frameworks underlying various data-driven methods, including latent variable methods, nonlinear methods utilizing kernel transformations, and dynamical systems methods, and (2) numerous case studies constructed to span the types of data encountered in chemical and biological process systems. The smart data analytics approach for fault detection is then demonstrated for a biopharmaceutical manufacturing process. As in past work [5], open-source software that implements the approach will be available for download.

[1] Chiang, Leo H., Evan L. Russell, and Richard D. Braatz. *Fault Detection and Diagnosis in Industrial Systems*. Springer, London, U.K., 2000.

[2] Isermann, Rolf. *Fault-diagnosis Systems: An Introduction from Fault Detection to Fault Tolerance*. Springer, London, U.K., 2005.

[3] Qin, S. Joe. Survey on data-driven industrial process monitoring and diagnosis. *Annual Reviews in Control*, 36(2), 220-234, 2012.

[4] Qin, S. Joe, and Leo H. Chiang. Advances and opportunities in machine learning for process data analytics. *Computers andamp; Chemical Engineering*, 126, 465-473, 2019.

[5] Sun, Weike, and Richard D. Braatz. Smart process analytics for predictive modeling. *Computers andamp; Chemical Engineering*, 144, 107134, 2021.

82a Implementation of a Small-Scale Capillary Shear Model to Facilitate Process Development of Disc Stack Centrifugation during the Harvest of Biopharmaceuticals

Steven Frey, Jane Liao, Stijn H.S. Koshari, Robert Luo, Antonio Ubiera

Harvest is generally the first step in the downstream processing of protein therapeutics and serves to clarify the cell broth by removing cells and debris. Disc stack centrifuges (DSCs) are often used during harvest, as they are capable of continuous operation and alleviate load on subsequent harvest filtration steps. It is important to understand how the DSC process parameters affect cell debris clearance and the degree of cell shearing, which can impact product quality attributes such as host cell protein and DNA concentration. However, as these DSCs require a relatively large amount of cell broth to operate, process development of the harvest step becomes impractical at the lab scale. Therefore, a small-scale model of the DSC is desirable but difficult to implement as DSCs impart shear forces on cells that differ from those generated by laboratory-scale centrifuges. Still, some small-scale DSC models have been proposed in literature. One such model uses a capillary shear device (CSD) to shear cell broth, which is then centrifuged with a lab-scale fixed-angle centrifuge to create DSC-like centrate. We have explored the use of this small-scale DSC model to process cell broth obtained from commercial-scale manufacturing batches of monoclonal antibodies. Specifically, we have evaluated how the model's parameters (flow rate of cell broth through the CSD, centrifugation time, and centrifugation speed) affect the characteristics of the model-generated centrate (turbidity, particle size distribution, and LDH activity – a measure of cell lysis). With a better understanding of how model parameters affect the model-generated centrate, the model can be used to replicate manufacturing-scale DSC centrate at the small scale such that harvest filtration studies can be conducted more efficiently. Further study of how the model's parameters relate to DSC parameters (bowl speed and flow rate) will ultimately allow us to evaluate if the model can be used to conduct process development of the entire harvest step (centrifugation and filtration) at the lab scale.

82b Using Particle Size Distributions to Predict the Filterability of Materials throughout the Processing of Monoclonal Antibodies

Steven Frey, Hong Zhang, Robert Luo, Antonio Ubiera

Throughout the downstream processing of monoclonal antibodies, filtration is used to remove particulates from various materials. As these materials can vary widely in their filterability, robust process development is necessary to prevent filter failure prior to reaching the desired filter loading at the manufacturing scale. Therefore, a method that can predict a material's filterability based on easily measurable material characteristics would benefit development and process transfer. For example, turbidity has previously been used as a rough measure to predict a material's filterability but is typically not an accurate predictor on its own. We explore the role of both turbidity and particle size distribution in the filtration of materials obtained from commercial-scale manufacturing batches of a monoclonal antibody. Ultimately, the results of this work may allow us to develop a model to predict a material's filterability based on its turbidity and particle size distribution.

82c Predictive CFD Mixing Model for Scale up/Down of Large Molecule Dug Product Formulation

Matthew Flamm, Joseph Rizzo, Hope Seybold, Tracey Mascaro, Avik Sarkar, Shweta Modi, Chandler Amato, Craig Ikeda

Large molecule drug product formulation typically requires mixing of multiple materials to form a consistent and uniform drug product before further processing. One common example is the mixing of multiple preformulated liquid streams to homogeneity in a mixing vessel. High viscosity products with high protein concentration and/or high excipient concentrations such as stabilizers represent a mixing challenge. Experiments at full scale manufacturing to determine operating ranges are costly and time consuming. Vessel geometries are also not always kept similar upon scale up or technical transfer of a process, which complicates design of scale down studies to determine suitable mixing parameters. CFD models have shown promise in delivering predictive tools [1] to shorten process development workflow and eliminate at scale experimental runs. However, there are fewer published reports in verifying these models over a wide range of tank scales and shapes, while quantifying accuracy and uncertainty.

A CFD model was built in MStar CFD to predict mixing time to homogeneity for multiple liquid streams. CFD predictions were verified against surrogate data at multiple process scales and tank geometries from ~10-1000L. Multiple tank geometry types (cylindrical and cubical tank shapes) were considered and will be compared to existing literature reports. Sensitivity of model inputs and the uncertainty in the model compared to experimental data will be discussed. The results of a qualified CFD model across relevant scales during process development is important for future use to reduce experimental runs at manufacturing scale.

[1] Thomas, J., Sinha, K., Shivkumar, G. et al. A CFD Digital Twin to Understand Miscible Fluid Blending. AAPS PharmSciTech 22, 91 (2021).

82d Model Aided Scale-up of Spray Drying for a Sticky API (low Glass Transition Temperature)

Ankur Kapil, Li Liu, Claire Wombell, Michael Hamlin

Spray drying can be used to generate very fine powders for inhalable drugs and to improve the bioavailability of solid dosage drugs using amorphous solid dispersion. However, spray drying is quite a challenging process to optimize and scale-up. There are many interacting parameters including drying gas flow rate, temperature, nozzle type, design, location, andamp; nozzle operating conditions (nozzle gas and liquid flow rate) that have a considerable impact on the product yield, residual solvent content, and the particle size and morphology. A traditional OFAT (One-factor-at-a-time), experiential, or a DOE based approach might not get an optimum due to the non-linearity and interactions between these parameters for Spray Drying and miss the detailed understanding from the mathematical model.

We have demonstrated a model-aided workflow along with the state of art process analytical characterization to optimize and transfer the process across three different Spray dryers for a sticky product. The material produced in this work has a very low Glass transition temperature resulting in yield loss due to the adhesion of the product to the wall. In this work, we model the droplet heat and mass transfer, droplet drying gas interactions, drying kinetics, and wall deposition for this sticky API. Mechanistic models (gFormulate®) and CFD models have been developed across three different equipment. gFormulate® model considers the droplet heat and mass transfer, drying kinetics, and heat loss from the wall. However, it doesn't consider the residence time distribution of droplets, wall deposition, agglomeration, andamp; coalescence in the Spray Dryer. gFormulate® models were used to estimate the sticky behaviour and the drying times based on the droplet size. CFD models have been developed using OpenFoam® to solve simultaneous heat and mass transfer between the droplets and the drying gas medium. CFD models were used to model the time to the wall for different droplet sizes, spray velocity, nozzle locations, distribution plate design, and drying gas velocity. Nozzle design and operating conditions have a significant impact on the process yield. The spray droplet size distribution is measured by Sympatec laser diffraction for the different nozzle designs and the operating conditions as an input to the model and experimental workflow. Mathematical models have been used to understand andamp; optimize the impact of operating conditions on the product yield across the scale. The understanding developed using the workflow has enabled the production of API with greater than 80% yield across the scales. The advantages and limitations of the mathematical models andamp; Spray Dryers across scales would be discussed in this presentation. Future work on the adhesion/cohesion of droplets in the Spray Dryer would be briefly discussed in this presentation.

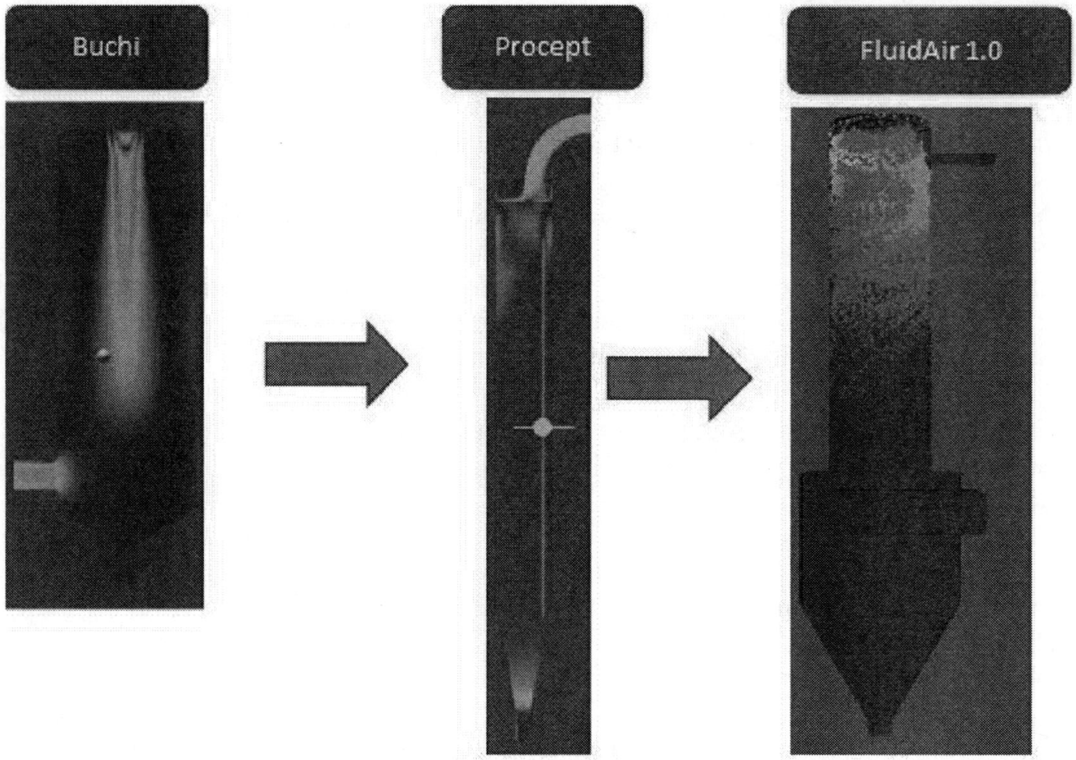
Buchi
Procept
FluidAir 1.0

82e In silico Modeling of Roller Compaction Processes for Scale-up and Tech Transfer: One Step Closer to Digital Twins

Elcin Icten Gencer, Dongying Shen, Duc Nguyen, David Perez-Aguilar, Mina Hamedi Rad, Shruti Gour, Saloni Daftardar, John Chung, Fernando Alvarez-Nunez, Fabrice Schlegel, Pablo A. Rolandi, Ananya Chowdhury, Maxwell Maritato, Rajarshi Sengupta, Killian Ryan

Dry granulation via roller compaction is a key unit operation in the manufacture of immediate release tablets. One of the challenges for late-stage development is identifying optimal roller compaction processing parameters for scale up through numerous Design of Experiment (DoE) studies, which require manufacturing time, drug substance and analysis. In this work, a first-principles in silico model available in gPROMS was extended to guide selection of key DoE studies and reduce the number of large-scale runs needed to identify optimal processing parameters. The prediction accuracy of this model was demonstrated for the scale up of the roller-compaction process at a commercial manufacturing site, with prediction errors within a few percent. This case study demonstrates the possibility of using this "VirtualPactor" model as a tool to reduce the number of experiments required to configure commercial roller-compaction process parameters. Instead of carrying out trial-and-error experiments to define the inputs for a large-scale roller compactor, product teams can carry out a DoE study on a small-scale roller compactor, use the data from the study to calibrate the VirtualPactor model, and use the calibrated model to find the inputs for the large-scale compactor that satisfy the output specifications. This model is developed into a web application for use by domain experts in an effort to democratize access to advanced modeling techniques and enable Quality by Design during process scale-up and tech transfer.

82f Optimizing the Design and Scale-up of Agitated Filter Dryer

Sravan Nallamothu, Guilherme Turezo, Mahesh Kadam, Ahmad Haidari

The drying operation is one of the most energy intensive unit operations and is often the rate limiting stage in the overall active pharmaceutical ingredient (API) manufacturing cycle. The optimization of operating conditions such as pressure, temperature, and mixing is crucial to avoid API degradation during the drying process. Agitated Filter Dryer (AFD) is widely used for its ability to conduct both filtration and drying operations in a single equipment, thus minimizing the product handling and operator exposure.

In this study, the Rocky DEM simulation tool is used to evaluate the particle dynamics of wet cake drying in an agitated filter dryer equipment. The impact of operating conditions and geometry configuration on drying is studied to optimize the design and operating space. Beyond that, the influence of various scale-up criteria on solvent drying behavior is also studied when going from a smaller lab scale dryer to a pilot scale dryer. Several important parameters such as type of solvent, particle size, particle shape, and solvent loading can be varied in a DEM simulation. The impact of these parameters on both solvent drying and the particle agglomeration is evaluated and can provide great insights for equipment design and operating conditions, leading to effective optimization of the AFD drying step during the API manufacturing cycle.

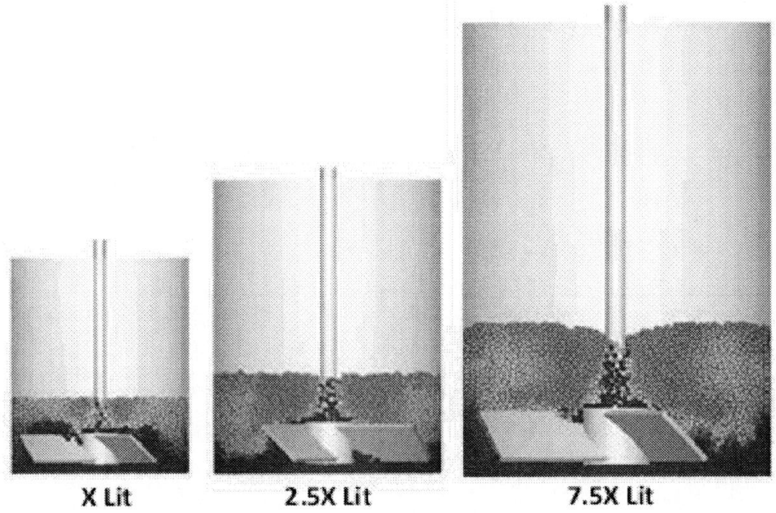

82g Scalability and Predictability of Terminal Wet Milling Processes

Carla Luciani, Mary Eccles, Michael Filios

Active pharmaceutical ingredients delivered through oral solid dosage forms usually require tight control of purity and particle size. For robust control and scaleup, practitioners are usually interested in correlating wet milling process parameters such as rotation rate, generator types, and number of batch turnovers to particle size and particle size distribution t [1-5]. High shear environments can also induce and accelerate form interconversion [6].

This work describes a process where particle size control and form control are achieved via high-shear rotor-stator wet milling. Practical considerations as well as mechanistic models used for process design, optimization, scale-up and overall control strategy will be discussed.

References

[1] C. Luciani, E. Conder, and K. Seibert. Org. Process Res. Dev. 2015, 19, 5, 582–589.

[2] A. Harter, L. Schenck, I. Lee, A. Cote. Org. Process Res. Dev. 2013, 17, 1335– 1344

[3] J. Engstrom, C. Wang, C. Lai, J. Sweeney. Int. J. Pharm. 2013, 456, 261– 268

[4] Luciani. Org. Process Res. Dev. 2018, 22, 9, 1328–1333

[5] K. Seibert, P. Collins, C. Luciani. Chapter 30: Milling Operations in the Pharmaceutical Industry. In Chemical Engineering in the Pharmaceutical Industry: Active Pharmaceutical Ingredients, Second Edition 2019. Editors: D. am Ende and M. am Ende, pp. 861-879.

[6] R. Carrasquillo-Flores, S. Wisniewski, V. Daftary, F. Lora-Gonzalez, T. Rosenbaum, R. Wethman, M. Huang, J. Wasylyk, J. Selekman, C. Choi, B. Mack, T. Razler, and J Engstrom. Org. Process Res. Dev. 2021, 25, 4, 1028–1035.

336a Continuous Flow Process Development for a Radical Chlorodifluoromethylation Reaction

Alexander Hesketh, Lu Han, Steven M. Guinness, Elaa Hilou, Ian Hotham, Robert Singer, Jordan Willie, Madison Gileau, Jason Ewers, Marcin Dembowski

A single step continuous manufacturing process has been developed to produce a drug substance intermediate via a radical fluoroalkylation[1]. The flow process has considerable advantages over the batch process, notably improved process safety concerns for a highly exothermic reaction with thermally unstable reactants. The flow process offers exquisite control of residence time at any scale, which limited the loss of product to hydrolysis during the reaction and work-up. Utilization of plate reactor technology and continuous liquid-liquid extractors resulted in a safe and highly efficient alternative to a batch synthesis. Further process optimization was completed, and a second-generation process was developed to avoid fouling issues, improve reactant efficiency and avoid dimethyl sulfoxide. The first- and second-generation processes were successfully scaled-up in a GMP facility at the multi-kilogram scale.

(1)

Meng, D.; Li, L.; Brown, A.; Desrosiers, J.-N.; Duan, S.; Hayward, C. M.; He, Z.; Hu, J.; Makowski, T.; Maloney, M.; Monfette, S.; Perfect, H.; Piper, J. L.; Zhou, M.; Widlicka, D. W. A Radical Chlorodifluoromethylation Protocol for Late-Stage Difluoromethylation and Its Application to an Oncology Candidate. *Cell Reports Physical Science* **2021**, *2* (4), 100394. https://doi.org/10.1016/j.xcrp.2021.100394.

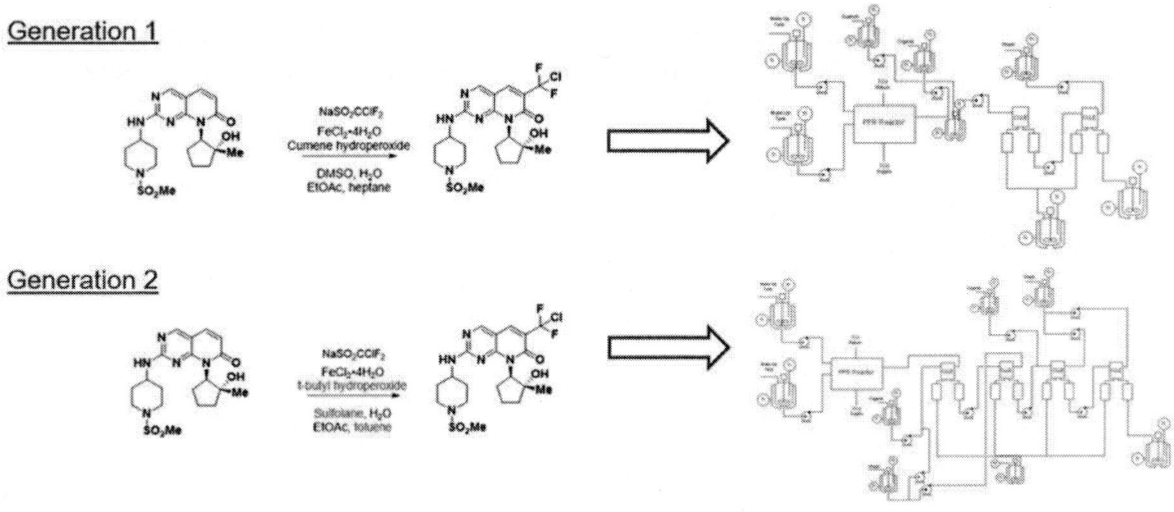

336b Flow Chemistry As an Enabling Tool for Scale-up of API Processes

Detian Gao, Tiago Vieira, Andrew Stevens, Nick Uhlig, Andrew Martins, Lars Heumann, Andrei Chtchemelinine, Pavel Badalov

Flow chemistry offers advantages to batch chemical processes that may have limitations in safety, heat transfer, and selectivity for the preparation of intermediates or active pharmaceutical ingredients (APIs). Herein we describe two case studies on the application of flow chemistry in process development and manufacturing.

Case 1: The use of flow chemistry to prepare an intermediate of remdesivir, an antiviral therapeutic approved for the treatment of COVID-19, includes a cyanation process that is demanding to perform as a batch process at production scale due to large quantities of hydrogen cyanide and cryogenic conditions. The reaction parameter control afforded by the flow process provides improved selectivity and yield. Additionally, the flow process and the flow reactor system design supported rapid scale-out in response to increasing production demand.

Case 2: Development of a flow chemistry process for the scale-up of an investigational drug candidate intermediate. A selective diisobutylaluminium hydride (DIBAL-H) reduction of a heterocyclic diester had acceptable selectivity and yield at laboratory scale, but lower regioselectivity, chemoselectivity, and yield were observed at pilot-plant scale under batch conditions. A flow chemistry process was developed with DoE that improved control of mass and heat transfer and provided better control of reaction time. This was quickly scaled up, allowing for the successful manufacturing of the intermediate at pilot-plant scale.

Reference:

1. Vieira, T.; Stevens, A. C.; Chtchemelinine, A.; Gao, D.; Badalov, P.; and Heumann, L.; Development of a Large-Scale Cyanation Process Using Continuous Flow Chemistry En Route to the Synthesis of Remdesivir *OPRD* **2020** *24* (10) 2113-2121
2. Uhlig, N.; Martins, A.; Gao, D.; Selective DIBAL-H Monoreduction of a Diester Using Continuous Flow Chemistry: From Benchtop to Kilo Lab *OPRD* **2020** *24* (10) 2326–233

336c Automated Continuous Crystallization Platform

Kakasaheb Nandiwale, Kevin Girard, Robert Pritchard, Charissa Nowak, Zheng Zhao, Steven M. Guinness

Continuous flow crystallization is an attractive mode of operation, due to its ability to generate consistent product quality while requiring a smaller footprint and lower production costs than its batch counterpart. We present the development of an automated continuous crystallization platform at Pfizer as a part of the Flexible API Supply Technology (FAST) initiative. This lab-scale continuous crystallization consists of a three-stage continuous stirred tank reactor (CSTR) system. LabVIEW™ software was designed and written in-house for the purpose of automated analysis and to control various hardware including a flow sonicator, feed pumps, and temperature controllers. We demonstrate the integration of a variety of in-line process analytical technology (PAT) tools, including Focused Beam Reflectance Measurement (FBRM), EasyViewer™, Blaze, and Infrared (IR) spectroscopy within this environment. By integrating equipment, PAT, and control software into a single, automated platform, enhanced real-time process understanding, identification of optimal conditions, and acceleration of development timelines are realized.

336d Developing Around Uncertainty in the Scale up and Tech Transfer of Flow Reactions with Highly Reactive Reagents

Robert Franklin, Travis Armiger, Pratiq Patel, Douglas Otte, Jackson Hall, Reed Larson, Val Rodrigues, Holst Halsey, Lisa Jellett, Nadine Kuhl, James P. Corry, Cheol Chung

Flow reactions involving highly energetic and moisture-sensitive compounds, like organolithiums and Grignard reagents, present unique challenges during scale-up and tech transfer for GMP intermediates. Concerns about operator safety and sample stability often prevent on-site release assays to confirm reagent concentrations, and small amounts of water in the equipment train can have a large impact on reaction performance. Additionally, differences in heat transfer and mixing performance across scales can cause uncertainty in the applicability of bench-scale process characterization experiments to full-scale operations. This talk will outline the strategies used to mitigate the uncertainties presented by the scale up of a multi-reaction flow step in the synthesis of the MK-1026 ketone intermediate, a penultimate step in the commercial route towards nemtabrutinib. As shown in the figure below, a batch deprotonation of the aryl bromide (ArBr) is carried out with the addition of methyllithium. A cascade of two reactions is then carried out in a tubular flow reactor, starting with the addition of *n*-butyllithium, followed by biaryl ether. An overcharge or undercharge of either of the two organolithium reagents can result in unacceptable impurity levels in the isolated ketone intermediate. A strategy for calculating, monitoring, and adjusting process stream flow rates enabled robust and consistent full-scale operation while maximizing yield. Comparison of repeated experiments across scales spanning multiple orders of magnitude enabled significantly more economical process characterization experiments on the bench to be used to guide large-scale processing decisions. The result was the successful transfer of the process from the Randamp;D labs and pilot plant to the final commercial site and demonstration of robust and consistent performance across eight full-scale batches over nine months of operation.

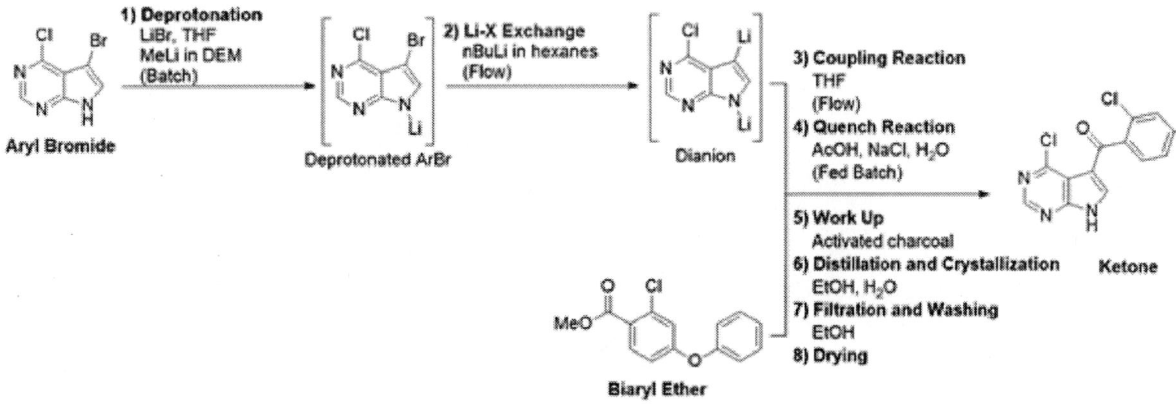

336e Distributed Pharmaceutical Manufacturing: The Quest for Pharmacy on Demand

Julian Chesterman, Luke Rogers, Tyler McQuade

Most generic pharmaceuticals sold in the United States of America are synthesized or formulated overseas, which leaves them vulnerable to supply chain disruptions. The World Health Organization (WHO) considers many of these generic pharmaceuticals to be essential for modern medicine. However, due to aging infrastructure and consolidated batch manufacturing processes, many of these generic pharmaceuticals appear on the FDA's drug shortage list on a regular basis. In particular, injectable drugs are prone to shortages and currently represent 109 of the 172 drugs on the shortage list.

As a result, we have developed a miniaturized technology platform for the synthesis, purification, and formulation of generic pharmaceuticals. The Pharmacy on Demand (PoD)™ platform enables domestic end-to-end manufacturing of the pharmaceutical supply chain from raw starting materials to formulated drug products, such as oral solid dosages or sterile injectables. By incorporating process analytical technology (PAT), flow chemistry, and automation into modular devices, a range of batch, semi-batch, and continuous processes can be conducted and scaled in an agile fashion depending on shifting market demands. The PoD platform is ideally suited for deployment using a distributed network of local or mobile manufacturing facilities to ensure a stable supply of quality medicines in both major population centers and disaster areas. By enabling cost-effective small-scale manufacturing, the PoD platform can replace aging centralized manufacturing facilities with a network of compact modern facilities to increase the reliability and responsiveness of the pharmaceutical supply chain.

336f Practical Challenges and Solutions to Continuous Grignard Chemistry

Sean Keenan, Ahmet Aloglu, Aibolat Koishybay, Stephen Born, Chuntian Hu

The Grignard reaction is one of the most powerful techniques for creating new C-C bonds and involves the formation of a highly reactive organometallic Grignard reagent. Despite its utility, Grignard chemistry presents significant safety hazards on large batch scales, such as runaway exotherms and magnesium handling concerns. Over the past several years, flow chemistry and continuous manufacturing methods have been applied to Grignard chemistry to create continuous modes that promote its utility while reducing the hazards.

Continuous operation of Grignard chemistry is not without its challenges, however. The development of an optimized Grignard reagent formation reaction necessitates careful consideration of many variables. For example, the heterogeneity of the Grignard reagent formation reaction requires special considerations for agitation and liquid separation. Creating a successful Grignard reagent formation process is further complicated by the inert and dry chemical condition requirements.

At CONTINUUS Pharmaceuticals, we have developed a laboratory-scale Grignard reagent platform to investigate the impacts of and address these challenges. The result is a workflow for process development, PAT model development, and scale-up. This workflow includes a method for converting a batch Grignard formation reaction to continuous on the 100 mL to 1 L scale, how to investigate activation and initiation techniques, practical IR model development strategies, common impurity forms to test for, Mg feed schedule considerations, among other topics. The CONTINUUS team has successfully used this workflow to generate a Grignard reagent at >95% yield throughout a multi-day campaign. Finally, the workflow informs a scale-up procedure with considerations for designing a 10 L commercial-scale Grignard reagent formation CSTR for a high potency API.

336g Enable the Implementation of Photochemical Reactions on Commercial Scale Via Using Continuous Flow Reactors

Hongwei Yang

Benzylic bromides represent an important class of building blocks in pharmaceutical industry due to the versatility in the synthesis of active pharmaceutical ingredients. The conventional thermal initiation approach for the synthesis of benzylic bromides via Wohl–Ziegler bromination presents several gaps from sustainability perspective, including the use of harmful chlorinated solvents and explosive radical initiator such as benzoyl peroxide or 2,2'-Azobis(2-methylpropionitrile). Here we present the use of photo chemistry to improve the greenness for this class of reactions by improving the yield and process safety, reducing the Process Mass Intensity (PMI), and eliminating the use of environmentally unfriendly solvents. We also demonstrate the use of continuous flow reactors to address the scaling up challenge associated with the decrease of light intensity along the irradiation path under batch mode. Via using the numbering up approach by the serial connection of continuous photo reaction units, we are able to achieve a productivity of hundreds of 150kg/day under GMP condition.

410b Combining Reaction with Crystallization and a Size-Based Biocatalyst Separation for the Continuous Manufacturing of Amoxicillin

Patrick Harris, Hossein Salami, Colton Lagerman, Ronald Rousseau, Martha Grover, Andreas Bommarius

Amoxicillin is a highly important drug used in the treatment of bacterial infections, and the largest produced antibiotic by volume globally. Its manufacturing has traditionally been conducted in a batchwise chemical synthetic process consisting of multiple steps, cryogenic refrigeration, along with the use of protecting groups and hazardous solvents. Research has been conducted in investigating the use of biocatalytic processing routes, which consist of a single reaction step, and are generally conducted in mild, aqueous conditions, generating much less waste and resulting in a fewer processing steps[1]. On the other hand, the biocatalyst, generally penicillin G acylase (PGA) from *E. coli*, not only catalyzes the one-step synthesis reaction, but also the hydrolysis of both the activated side-chain donor 4-hydroxyphenylglycine methyl ester (4-HPGME) as well as the product amoxicillin. Several strategies have been investigated to limit these side reactions in similar drugs, such as in situ product removal via complexation[2], aqueous two-phase systems[3], supersaturation of substrates[4], and more. While these strategies offer improvements over a standard one-pot aqueous system, they either increase difficulty in downstream processing steps or are only applicable to batch systems. Nevertheless, many improvements to the biocatalytic processing of amoxicillin are still required to make it feasible on an industrial scale.

In this work, we combine the reaction and crystallization in the same mixed suspension mixed product removal (MSMPR) vessel to improve the selectivity and productivity of the system as well as intensify the overall process. Additionally, we operate the process in a continuous fashion, leading to increased productivity, decreased process footprint, and easier operation in comparison to a batch process. Biocatalyst retention within the reactor was achieved by immobilizing PGA on a solid support of a defined size range (300-425 μm), and withdrawal of the product slurry through a stainless-steel mesh filter of 300 μm, thereby achieving a size-based separation between product crystals and biocatalyst. As amoxicillin trihydrate crystals will grow much larger than 300 μm normally without intervention, an external wet-mill loop was employed to decrease their size below the mesh filter threshold. Additionally, crystallization of amoxicillin was observed to be rate limiting, so we investigated the use of a second MSMPR, operated in the absence of biocatalyst, to allow for its complete desupersaturation from solution. It was determined that a residence time of 1 hour was sufficient to desupersaturate amoxicillin and improve product yield by 10%. Lastly, while these process improvements led to a high productivity of 500 g/L/d, the conversion of the beta-lactam donating moiety, 6-aminopenicillinoic acid (6-APA), and the side-chain donor, 4-HPGME, were low at only 53% and 40%, respectively. To address this limitation, we were able to isolate both substrates separately via pH-swing crystallization at pH values of 4.0 and 8.0 for 6-APA and 4-HPGME, respectively, exploiting their solubility's opposite sensitivity to changes in pH. With the recovery and reuse of substrates, the overall conversion, defined as the fraction of the compound fed into the system which does not leave the system as waste, was improved to greater than 90% for both substrates.

1. McDonald, M.A., A.S. Bommarius, and R.W. Rousseau, *Enzymatic reactive crystallization for improving ampicillin synthesis.* Chemical Engineering Science, 2017. **165**: p. 81-88.
2. Li, D., et al., *Enhanced enzymatic production of cephalexin at high substrate concentration with in situ product removal by complexation.* Food Technology and Biotechnology, 2008. **46**: p. 461+.

3. Aguirre, C., et al., *Partition and substrate concentration effect in the enzymatic synthesis of cephalexin in aqueous two-phase systems.* Process Biochemistry, 2010. **45**(7): p. 1163-1167.
4. Youshko, M.I., et al., *Penicillin acylase-catalyzed synthesis of ampicillin in "aqueous solution–precipitate" systems. High substrate concentration and supersaturation effect.* Journal of Molecular Catalysis B: Enzymatic, 2000. **10**(5): p. 509-515.

410c Moisture Content Monitoring in Continuous Drug Substance Isolation Manufacturing Platform

Inyoung Hur, Daniel Casas Orozco, Zoltan Nagy

With the shift of paradigm from batch to continuous in pharmaceutical manufacturing, integrated filtration-drying has become an important emerging technology in continuous drug purification to produce high-purity small molecule oral dosage drugs [1]. Among several alternatives for the intensified filtration unit, the continuous filtration carousel (CFC) has become an efficient technology for handling various production scales following crystallization [2]. The CFC's main body is a cylindrical carousel, equipped with stations, which are each allocated to one or more steps including filtration, washing, drying, and material transfer and discharge. The port rotates to the next station every predefined cycle time, allowing a sequence of small batch processes to handle continuous separation of solid forms of the active pharmaceutical ingredient (API). By adopting separated small-scale batch process within the continuous manufacturing (CM) horizon time, the extensive knowledge gained previously on batch filtration and drying can be easily adapted in to the integrated framework : 1) their lack of steady states and unknown model parameters in filtration-drying step, 2) dynamic critical material attributes (CMAs), namely feed properties of the slurry coming from the upstream crystallization step.

Quality-by-design (QbD) initiatives have been developed in the last years to overcome the challenges posed by input variability and model parametric uncertainty, and thus expedite the modernization of the pharmaceutical process development in the CM [3]. QbD defines the viable operating region of given sets of CMAs and critical process parameters (CPPs) to fulfill the target quality of critical quality attributes (CQAs). With the increasing availability of process analytical tools (PAT) and digital design tools, QbD has been used in various open-loop studies that create and use process digital twins to optimize process metrics (CQAs, costs, waste generation) with CPPs as decision variables [4, 5]. For the pharmaceutical isolation step, crystal properties and final product impurities of regulated materials are the main CQAs. Since impurities remaining in the solid at the end of the drying step will be directed to the downstream process, failure to reach the drug's targeted purity will lead to off-spec products under the stringent regulatory screenings on the CQAs [6]. Furthermore, the dynamics of crystal properties and final product purity from the simulation are not always in sufficient agreement with real process data. This plant-model mismatch may come from the presence of gross measurement noise, such as outliers and drifts, and the fact that not all model states can be measured, which makes model calibration harder. Improving the estimation and monitoring of the CQAs is critical to incorporate the decision-making framework developed from the design space (DS) into the closed-loop control of the product quality to a higher level of assurance.

We have developed a moving horizon estimation (MHE) framework that relies on state estimation of average moisture content of API filtration cakes, and applied it in the pharmaceutical crystallization-filtration-drying step of a real continuous drug substance isolation line. The MHE framework is based on a first-principles mathematical model describing solid-liquid separation of porous media consisting of crystal feed from an upstream crystallization unit. Since main CQAs in the analyzed case study, namely, cake average moisture and API productivity, cannot be measured in real time, state estimator was critical for their accurate prediction and further model recalibration by the MHE framework. Additionally, offline analysis of the final product CQAs was carried out to validate the proposed MHE workflow. Considering the non-negligible computational time required to solve both the model equations and the MHE problem, a sensitivity analysis was performed to rank uncertain model parameters from the system to moisture content of the final product. The analysis shows that from the steps carried out on the continuous drug isolation operations, filtration and drying model parameters contribute the most to changes in moisture content. Hence, the MHE problem was formulated focusing on the filtration and dryer steps in order to minimize computational time. To correct the model prediction due to the accumulation of the error within the operation timeline, the MHE estimates the moisture content from carousel using a digital twin of continuous crystallizer-CFC and measurements of the inlet and outlet temperature of dryer, pressure drop during the processes, and dry gas outlet composition. From this we estimate the minimum cycle time for filtration and drying to satisfy the target purity constraint, so that the productivity of the system can be improved. The crystallization step is accounted by a calibrated mathematical model, and it enables the inline calculation of the dynamic characteristic cake property capturing various range of filterability and drying dynamics. In this work, the OPC communication protocol was utilized to access to real-time sensor data, and the Python-based, object-oriented platform, PharmaPy [7], has been used to solve the CFC model equations and the MHE problem. The proposed inferential monitoring scheme was evaluated with open-loop and closed-loop experimental data and then integrated into a continuous crystallizer-CFC process. The performance of the proposed monitoring framework was successfully demonstrated by reconstructing the moisture content of the final product by the estimator in the presence of gross measurement errors.

References

[1] Price, C. J., Barton, A., andamp; Coleman, S. J. (2020). CHAPTER 13. Continuous Isolation of Active Pharmaceutical Ingredients. In The Handbook of Continuous Crystallization. The Royal Society of Chemistry. https://doi.org/10.1039/9781788013581-00469

[2] Destro, F., Hur, I., Wang, V., Abdi, M., Feng, X., Wood, E., ... andamp; Nagy, Z. K. (2021). Mathematical modeling and digital design of an intensified filtration-washing-drying unit for pharmaceutical continuous manufacturing. Chemical Engineering Science, 244, 116803.

[3] Lee, S. L., O'Connor, T. F., Yang, X., Cruz, C. N., Chatterjee, S., Madurawe, R. D., Moore, C. M. V., Yu, L. X., andamp; Woodcock, J. (2015). Modernizing Pharmaceutical Manufacturing: from Batch to Continuous Production. Journal of Pharmaceutical Innovation, 10(3), 191–199. https://doi.org/10.1007/s12247-015-9215-8

[4] D. Casas-Orozco *et al., "Digital Design of a Lomustine Manufacturing Process Using Pharmapy", in AICHE Annual Meeting, 2021*

[5] Su, Q., Bommireddy, Y., Shah, Y., Ganesh, S., Moreno, M., Liu, J., Gonzalez, M., Yazdanpanah, N., andamp; Connor, T. O. (2019). Data reconciliation in the Quality-by-Design (QbD) implementation of pharmaceutical continuous tablet manufacturing. International Journal of Pharmaceutics, 563(April), 259–272. https://doi.org/10.1016/j.ijpharm.2019.04.003

[6] International Council for Harmonisation, 2016. Impurities: guideline for residual solvents Q3C(R6)

[7] Casas-Orozco, D., Laky, D., Wang, V., Abdi, M., Feng, X., Wood, E., Laird, C., Reklaitis, G. V., andamp; Nagy, Z. K. (2021). PharmaPy: An object-oriented tool for the development of hybrid pharmaceutical flowsheets. Computers and Chemical Engineering, 153, 107408. https://doi.org/10.1016/j.compchemeng.2021.107408

410d Design of a Pilot Plant for Integrated Continuous Manufacturing of a Steroid API

Guanghui Zhu, Andriy Neshchadin, Zachary Fejedelem, Janaka Gamekkanda, Ahmet Aloglu, Sean Keenan, Stephen Born, Chuntian Hu

Continuous manufacturing (CM) offers advantages in improving process safety and quality control, reducing costs and lead time, and providing access to additional design space compared to batch processes. Unlike batch process development, where development work can be accomplished with a small set of equipment at a given scale (i.e. one jacketed reactor for reaction, extraction, distillation, and crystallization during lab development), CM process development usually requires unique equipment for each unit operation. During scale-up, this set of equipment needs to be redesigned to accommodate the increased throughput.

At CONTINUUS Pharmaceuticals, we developed individual unit operations for a CM process in small scale lab equipment. In order to demonstrate scale-up performance and integrated process performance, we piloted the process with a throughput between lab and anticipated commercial scale. To accommodate the project timeline, design and construction of the new pilot plant needed to occur in parallel with lab development, so that the pilot plant would be operational immediately when lab development was complete. This posed many unique challenges, namely that the equipment had to be designed early with limited process knowledge to accommodate the project timelines given the supply chain shortages and long lead times. We developed an uncertainty-based calculation for process parameters that informed the pilot plant design which evolved with improved process understanding acquired from lab development. The equipment was designed to cover 95% confidence interval on the process variability.

The detailed engineering design of the pilot plant retrospectively influenced the concurrent lab development activities by dictating scale-up constraints. This led to substantial process changes but provided sufficient time to de-bottleneck the process to integrate different unit operations into an end-to-end process. Ultimately, performing lab development and pilot plant design concurrently resulted in an unexpected synergistic benefit.

In conclusion, we designed and constructed a pilot plant for an integrated CM process. We quantified uncertainties during equipment design and made data-driven decisions to drive the project timeline. Executing lab development and pilot plant design concurrently ensured a smooth transition during scale-up, process integration, and tech transfer.

410e Continuous Twin-Screw Wet Granulation and Fluid-Bed Drying at the Pilot Scale and Comparison to Batch Granulation and Drying

Seth Forster, Erin Dippold, Abbe Haser, Imre Homolya, Daniel Emanuele, Robin Meier

Continuous manufacturing promises many advantages such as accelerated development and commercialization, flexible batch sizes, improved process monitoring and control, and process intensification. Many pharmaceutical oral solid dosage processes have been adapted to continuous operation, though continuous wet granulation has had limited application thus far due to the challenge of truly continuous drying.

L.B. Bohle has developed a novel, continuous vibratory fluid-bed dryer system at both pilot (QbCon® 1) and commercial (QbCon® 25) scales. The systems are comprised of an integrated loss-in-weight feeder, a twin-screw granulator with liquid injection, the continuous dryer, and discharge with the option for integrated process analytical tools throughout.

This study sought to demonstrate the feasibility of continuous wet granulation and continuous drying at the pilot scale (1 – 3 kg/h) using placebo and active formulations across a range of process conditions. Each of the formulations had been previously used for batch high-shear or fluid-bed granulation.

The QbCon® 1 system enabled much more rapid evaluation of process parameters and reduced down-time between conditions relative to batch processes. Conditions were altered and representative granules were collected in a few minutes. Twin-screw wet granulation required lower granulation fluid levels (GFL, also referred to as liquid/solid ratio, L/S) than for the batch operations, probably due to improved distribution of the granulating fluid. The vibratory fluid-bed dryer dried the granules to less than 2% loss on drying in a few minutes, at up to 30% GFL. Little drying process optimization was required. Granule size distribution, flow behavior, and tablettability performance indicate that the QbCon® 1 system can produce granules with comparable properties to the batch processes.

These results suggest that the granulator and novel vibratory fluid-bed dryer can effectively process a range of formulations and continuously produce granules with the desired physical properties.

QbCon® 1

410f Developing a Process Analytical Technology for Particle Size Determination in Twin Screw Granulation Using Acoustic Emissions

Hassan Abdulhussain, Michael R. Thompson

PURPOSE: Twin screw granulation (TSG) produces a particle size distribution (PSD) that is complicated to monitor due to its bimodal nature. Acoustic emission (AE) offers significant potential for robust monitoring since impacts can be related to particle size, making it a suitable approach for developing a process analytical technology (PAT) for TSG. To predict weight fractions for a broad range of anticipated granule sizes, it is postulated the intensity of a set of frequencies in an acquired AE spectrum can be correlated to the impact mechanics of specific particle sizes. The critical challenge in developing this technique involves compensating for the differences in AE intensity between small and large particles in order to come up with an accurate accounting of impacts. This phenomenon where some sounds are overshadowed by others is known as auditory masking.

METHODS: Flowlac 100 lactose monohydrate powder (Meggle GmbH) was granulated in a 27 mm 40 L/D co-rotating twin screw extruder (Leistritz Advanced Technologies Corp.) with METHOCEL E3PLV (International Flavour and Fragrances Inc.) as the aqueous binder. An stainless steel impact plate was placed 20 cm below the exit of the extruder with an attached broadband AE sensor (Mistras Group) to record particle impacts. Multiple setup configurations for the plate were examined, looking at thickness (0.7, 1.0, and 1.5 mm) and angle of inclination (10-60°), with particle collisions recorded with a FASTCAM SA-Z type 2100k high speed camera (Photron Limited) at 20,000 frames per second to arrive at an optimal system.

The optimal acoustic sensor-plate configuration was used to record particle collisions for granules exiting the extruder at liquid-to-solid (L/S) ratios of 8-14% to produce PSD with differing amounts of fines (<300 μm), coarse particles (>2000 μm), and suitable sizes for tableting (600-1500 μm). The collected granules were characterized using a sieve shaker to determine the weight fractions of the PSD from 49 to 7150 μm. The AE signal was pre-processed by applying a discreet Haar wavelet filter followed by a fast Fourier transform to view the signal in the frequency domain. For the wide range of particles dropping onto the plate, auditory or spectral masking complicated the analysis by the model. To address this issue, a digital impact signal filter was conceived to alleviate auditory masking at the impact plate, derived from Hertz theory for single particle impacts.

An artificial intelligence neural network (ANN) model was created to relate the frequency domain spectrum to a PSD. The model architecture consisted of one input layer, three hidden layers, followed by one output layer. The model had activation functions of the Rectified Linear Unit for the input and first hidden layer. The second and third hidden layers had the hyperbolic tangent and sigmoid functions, respectively, while the output layer had a linear activation function. The root mean squared error (RMSE) was used to quantify the prediction error. Data from the multiple experimental trials with the extruder for each L/S ratio, were split into 80% training and 20% testing sets. Stratified splitting was employed to ensure the training and testing sets contained an equal proportion of data for each L/S ratio.

RESULTS: Trials examining the effect of inclination angle for the plate found that 60° (relative to the horizontal) was ideal. At angles below 45°, the AE signal should have been stronger due to a higher normal force attributed to a collision; however, due to mass accumulating on the surface of the plate, the signal was actually dampened. High speed camera images showed a significant amount of granular damage at 45°, associated with the normal force of collision. At 60° the decreased normal force in particle impacts reduced the damage observed and the granules departed the collision site by sliding/rolling with little detectable secondary impacts.

Increasing the plate thickness reduced the signal amplitude but did not affect the frequency information within the signal. At lower thicknesses, particle impacts caused detectable vibrations of the plate itself and seemed to increase the likelihood of secondary collisions by the granules. Increasing the thickness to 1.5

mm reduced these vibrations, making it the preferred thickness for the setup.

Prediction errors for the ANN model showed that use of the digital impact signal filter significantly improved the accuracy, from a maximum deviation of 4.2 wt% to 1 wt% for all particle sizes. Much of the error was localized around the particle size fraction at 2230 μm. After the digital impact signal filter was applied to the AE data, the loss error was no longer localized around 2230 μm, highlighting the negative effect that auditory masking was having on the AE signal in the original model. Diminishing auditory masking of the signal allowed the model to more equally consider the contribution of all frequencies (and their related particle collisions) in the predictions.

CONCLUSIONS: A robust approach for PAT based on ultrasonic acoustic emissions has been successfully found to monitor the complex bimodal particle size distribution attributed to continuous twin-screw granulation. This preliminary study for the AE-based PAT approach was trained with lactose monohydrate, whereas future studies will report on the relevance of different excipients and their individual nature of hydration on the acoustic emissions of impacting granules.

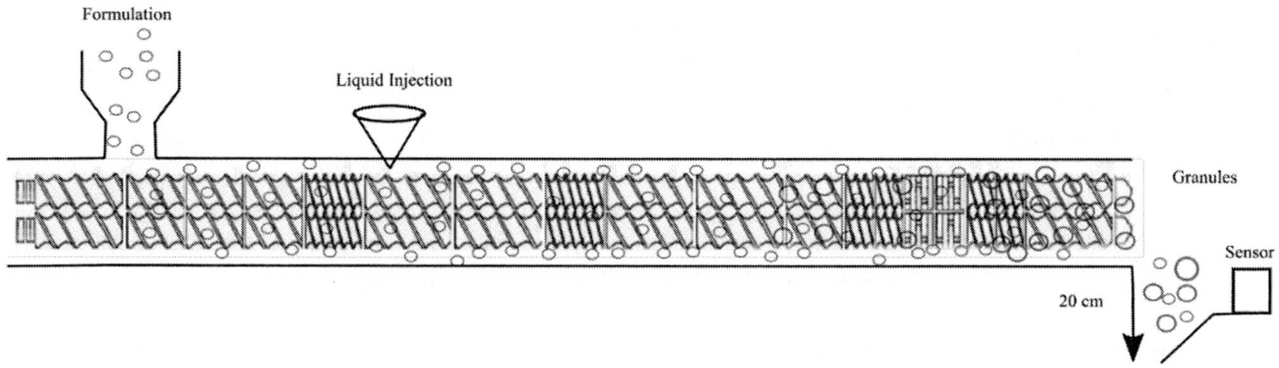

410g Computational Modeling of the Residence Time Distribution of Twin Screw Extrusion for Continuous Manufacturing Applications

Nisha Raman, William Blincoe, Morgan Giles, Nathan Rudd, Athanas Koynov

Twin screw extruders (TSE), where two co- (or counter-) rotating screws pump viscous material along the length of the barrel, while providing both distributive and dispersive mixing, have found many applications across multiple industries – polymer, food, pharma, etc. As an inherently continuous unit operation, TSEs can be seamlessly integrated into any continuous drug product manufacturing train to provide, e.g. high-shear wet granulation, melt granulation, amorphous solid dispersion hot-melt extrusion, etc. Optimization of the performance of the TSE relies on the control of the residence time distribution (RTD) of the material in the extruder.

In this work, the RTD of a TSE was captured by leveraging near infrared (NIR) process analytical technology (PAT) positioned at the die of the extruder. Following pulse additions of a NIR-detectable tracer to the material fed to the TSE, NIR spectra of the polymer extrudate was captured in-line and was translated to percent tracer content. In-line tracer concentration data was then used to model the RTD of the TSE using a tank-in-series model. Additionally, a plug flow time-delay component was convolved with the tank in series model to more closely match the measured distribution. Finally, the RTD was modeled for different screw speeds and feed rates to compare process attributes as a function of extruder fill level. Using the established models, we aim to predict mixing behavior in the TSE for varied total throughputs and leverage this information to engineer a control system for maintaining a uniform tracer concentration in the extrudate during scale up to a commercial level.

139a Development of a Single-Tablet-Scale Direct Compression Process for the on-Demand Production of Personalized Tablets

Andreas Kottlan, Andrea Zirkl, Jakob Geistlinger, Jörg Prassl, Johannes Obersriebnig, Benjamin Glasser, Johannes G. Khinast

Our present work aims to address two challenges the pharmaceutical industry is currently facing. First, sensitivity of the complex global supply chains which can lead to shortages of important drug products, if one link in the chain fails. Second, meeting the demand for personalization of pharmaceutical treatments with respect to genetics, age, sex, bodyweight, etc. Both challenges are connected to the fact that most tablets are currently produced in high volumes of fixed dose units unspecific to the demands of individual patients and by processes fine-tuned to maximize efficiency and to minimize cost in this competitive market. A solution can be producing tablets on demand in the needed quantity and composition for a patient in a process minimizing the number of unit operations as well as the necessary number of ingredients. Pursuing this solution, we developed a direct compression process consisting of a gravimetric dosing, blending, and compaction step producing one tablet at a time. This process concept circumvents common challenges in direct compression like consistency of flow in die filling, segregation of the blend, and content variability in the final tablet while allowing to adjust the dose in each tablet. To arrive at a reasonable production rate, innovative technologies for the dosing and mixing of small amounts of powder were found and tuned to the requirements of the envisioned process.

Dosing was realized using a device based on a volumetric microfeeder (*1*) which was combined with a weighing cell to provide gravimetric feedback. This facilitated the implementation of a control strategy to enable dosing of various pharmaceutical powders in the mg-range. The vessel used to collect the powder in the dosing step is subsequently sealed and used as the mixing container and finally as compaction die. Consequently, the powder masses dosed in the first step are exactly the amounts found in the final dosage forms, as there is no powder transfer or equipment entering or leaving the vessel. Therefore, blending needs to take place inside a sealed vessel having the dimensions of a compaction die (i.e. 10 mm in diameter and 20 mm in height). To solve this task, we used vibratory mixing as this was identified as a promising approach to blend small amounts of pharmaceutical cellulose powders to a satisfactory degree of homogeneity within less than 3 seconds (*2*). Compaction was done using a single punch tablet press consisting of an electric linear actuator equipped with a load cell and a mechanism to separate the tablet from the die.

To validate the capability of this process to produce high quality tablets, first each process step was investigated isolated. To do so, the dosing and mixing processes were investigated using different excipient and API powders evaluating dosing performance and blend homogeneity qualitatively and quantitatively. Finally, model tablets containing 200 mg of API (Ibuprofen), 190 mg of microcrystalline cellulose, and 10 mg of a disintegrant (crospovidone) were produced using the developed setup. These tablets were analyzed for their Critical Quality Attributes (CQAs), namely the API content, tensile strength, and dissolution profile.

The results regarding the dosing performance show the capability of dosing various powders largely independent of their flow characteristics within a variability of ±10% for masses down to 20 mg. Below that dose, cohesion and particle size decrease the dosing performance. Concerning the mixing process, vibration parameters were found allowing to blend different systems consisting of one coarse material (d_{50} > 100 µm) and a fine material (d_{50} < 50 µm) or two coarse materials within 5 seconds to a satisfactory homogeneous state. Challenges were identified for systems consisting solely of fine material. Finally, the investigations on the produced Ibuprofen tablets revealed a mean API content of 101.21 ± 0.93 % of the desired amount and a tensile strength of 1.00 ± 0.04 MPa. In dissolution testing 87.2 ± 3.3 % of API were in solution within 15 minutes meeting the USP criterion for immediate release Ibuprofen tablets (80 % in 60 min (*3*)).

We therefore conclude that the developed setup is a valuable platform for manufacturing personalized tablets and to produce tablets on demand from a limited number of ingredients. Future work will focus on resolving limitations of dosing as well as in blending of cohesive powders. Furthermore, additional model systems will be tested to investigate the operation space of the developed platform.

References

1. M. O. Besenhard *et al.,* Micro-feeding and dosing of powders via a small-scale powder pump. *International journal of pharmaceutics*. **519**, 314–322 (2017), doi:10.1016/j.ijpharm.2016.12.029.
2. A. Kottlan, B. J. Glasser, J. G. Khinast, Vibratory mixing of pharmaceutical powders on a single-tablet-scale. *Powder Technology*. **387**, 385–395 (2021), doi:10.1016/j.powtec.2021.04.040.
3. USP, *Ibuprofen Tablets. In: USP29-NF24* .

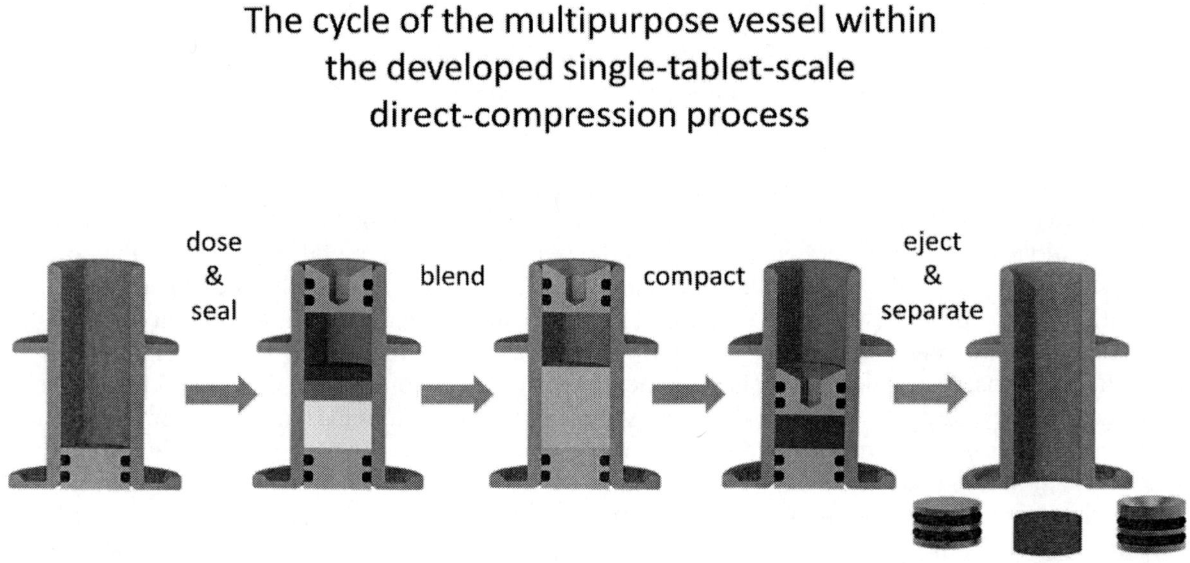

139b On the Applications of Pervaporation in the Pharmaceutical Industry

Akshay Korde, Sharad Maheshwari, Manish Kelkar, Shashank Shekhar, Ryan Ellis, Moussa Boukerche, Sarah Co, Eric Moschetta, Harsit Patel, Nandkishor Nere

Pervaporation is a tried and tested membrane technology in the chemical industry. However, the applications of pervaporation have been seldomly practiced in the pharmaceutical industry.

The use of either hydrophilic or hydrophobic membranes in pervaporation can allow for the selective removal of solvent/s from a mixture of solvents. Several unit processes in the manufacturing of active pharmaceutical ingredients (APIs) are sensitive to the presence of water, and hence removal of water is a ubiquitous operation in the pharmaceutical industry. It is typically achieved through distillations, commonly known as chase or azeo-distillations. These distillations could use exorbitant amounts of organic solvents to achieve the desired water content. Hydrophilic pervaporation membranes that can selectively remove water out of a solution, can be very effective at addressing this problem.

In this work, we will present case studies depicting use of pervaporation to optimize several standard unit operations in the pharmaceutical industry such as chemical reaction, distillation, and crystallization. We will highlight how pervaporation can help to design alternate processes that are greener and robust.

Disclosure:

Akshay Korde, Sharad Maheshwari, Manish Kelkar, Shashank Shekhar, Ryan Ellis, Sarah Co, Eric Moschetta, Harsit Patel, and Nandkishor Nere are employees of AbbVie. Moussa Boukerche is a former AbbVie employee. All authors may own AbbVie stock. AbbVie sponsored and funded the study; contributed to the design; participated in the collection, analysis, and interpretation of data, and in writing, reviewing, and approval of the final publication.

139c Designing Rules for Continuous Separation of Enantiomers By Orthogonal Electrochromatography

Michal Pribyl

The separation of enantiomers in a continuous manner represents a nontrivial task. Identical physical properties of an enantiomeric pair render the standard separation methods unusable, and its separation traditionally relies on the use of a chiral environment. Here, we focus on the method of continuous steady-state separation of enantiomers of mandelic acid in a chromatography column packed by a chiral sorbent. We consider the orthogonal orientation of the pressure and electric field gradients, i.e., the driving forces of the separation process. The transport of solutes in a mobile phase and a porous stationary phase is analyzed. We describe the behavior of mandelic acid inside porous particles by introducing a pseudo-homogeneous model in which the solute binding to the solid phase effectively decreases the diffusivity of the solute. For simplicity, the mandelic acid partitioning between the liquid and solid phase was described by a linear isotherm. However, the mathematical model can be modified using nonlinear adsorption isotherms. The model only requires the knowledge about the equilibrium relationship between the solute concentrations in the solid and liquid phases. Two local balances coupled through the overall mass transport coefficient model the mandelic acid transport inside both pseudo-homogeneous environment and mobile phase. The model equations are numerically analyzed in spatially two- and three-dimensional domains and the separation efficiency of mandelic acid is determined in the parametric space of the control and other parameters. More importantly, we derived simplified algebraic expressions determining the concentration trajectories of mandelic acid under the assumptions of phase equilibrium and neglected diffusion/dispersion transport. These expressions may assist in optimizing the geometry of an orthogonally operating electrochromatographic device and setting the control parameters such as the applied voltage or pressure difference. The predictions based on the simplified expressions are in good agreement with both available experimental observations and numerical simulations of the full model. Finally, we show that orthogonal electrochromatography can separate enantiomers continuously in a steady-state regime under optimized operating conditions. The separation efficiency strongly depends on the enantioselectivity of the stationary phase. Our results offer basic designing rules for orthogonal electrochromatographic devices that can be integrated in continuously operating micro- or milli-plants for the continuous synthesis and separation of pharmaceuticals and other special chemicals.

References

- Přibyl, M.; Izák, P.; Slouka, Z.: A mathematical model of a lateral electrochromatography device for continuous chiral separation, *Separation and Purification Technology* **2022**, *282*, 120033.

Acknowledgement: The author thanks for the support by the grant of the Czech Science Foundation (grant no. 20-09980S).

139d Thermal Analysis of the Wet Stirred Media Milling Process Used for the Production of Drug Nanoparticle Suspensions

Gulenay Guner, Natasha Seetharaman, Sherif Elashri, Mirsad Mehaj, Ecevit Bilgili

Nanocomposite formation has been of the most popular platform approaches in dissolution enhancement of poorly water-soluble drugs as the higher surface area and mass transfer coefficient of the nanoparticles can lead to faster dissolution than their micron-sized counterparts [1]. Wet stirred media milling (WSMM) has been the most preferred method for the manufacture and preparation of marketed drug nanocrystal-based products and nanocrystal suspensions in scientific studies [2]. The reason for this high demand for WSMM is that it has the potential of producing high drug-loaded, stable nanosuspensions while being a reproducible, scalable, solvent-free, and environmentally benign process [3]. On the other hand, if the process conditions are not selected carefully, WSMM can suffer from long cycle times, high energy consumption, and overheating problems [4].

In WSMM, like any other milling operation, only a small fraction of the mechanical energy spent on mixing the slurry is used to deform the drug particles [5]; most is eventually converted into heat through dissipative processes (viscous losses, inelastic bead collisions, etc.), which causes the temperature to rise during the milling. The occurrence of notable temperature rise required researchers to shut down the mill intermittently and cooled it without further milling [6, 7]. Nonetheless, temperature plays an important role and needs to be controlled during the WSMM of drug suspensions. First, the physical or chemical stability of a drug suspension could be susceptible to temperature [8, 9]. Drugs especially thermolabile drugs require stringent low-temperature control to avoid chemical degradation. Moreover, a polymer like hydroxypropyl cellulose (HPC), which is a commonly used steric stabilizer in drug nanosuspensions, has a lower critical solution temperature (LCST), a.k.a. clouding point temperature, of ~45 oC [10]. HPC may start precipitating out of aqueous solution at or above LCST, which may in turn cause physical destabilization of drug nanosuspensions. Also, a higher temperature during WSMM or storage could facilitate Ostwald ripening and growth of drug nanoparticles [11, 12]. In many industries where WSMM is used frequently such as ores, inkjet inks, ceramics, etc., a stringent low-temperature control is not needed as these materials do not have a risk of thermal degradation during the WSMM. In fact, for these materials, higher temperatures are likely to be tolerated as the viscosity and associated power consumption during the milling are lowered at higher temperatures. On the other hand, in the pharmaceutical industry, a stringent temperature control during the WSMM is required for the reasons mentioned above. Even though temperature control and heat dissipation are important for a WSMM process, it is interesting to note that the impact of process parameters on heat dissipation and temperature during the WSMM has not been investigated. We hypothesize that temperature is a strong function of the power consumption during the WSMM as observed during our previous study [13].

This study aims to examine the timewise evolution of the mill outlet temperature and chiller temperature as a function of intermittent milling and cooling time (if mill is shut down). Fenofibrate was used as the model drug and its stable suspension with HPC was prepared. The precursor suspensions were milled at three different stirrer speeds, bead loadings and bead sizes to explore their impact on the timewise evolution of temperature at the mill outlet. We aimed to keep the mill outlet temperature below 45 oC. The mill was shot down with continued cooling whenever the outlet temperature reached 45 oC, which led to multiple intermittent milling/cooling cycles. The fenofibrate particles were subjected to a total of 60 min milling; however, the actual cycle time was longer when multiple intermittent milling cycles followed by cooling were applied. The temperatures at the mill outlet and the chiller are recorded every minute or every 30 s, depending on the heat release rate (based on knowledge from exploratory experiments). After milling, the particle sizes were measured using laser diffraction. Also, densities and viscosities are determined to be used in Reynolds number calculations, which will be used for explaining the power–heat generation relationship. In order to identify any quantitative trend between the process conditions and the heat generation, specific times and temperatures were defined such as the maximum temperature, the temperature at 5 min of milling, and the time required to reach 25 oC, etc.

The figure shows examples of the temperature profile to give an overall idea, where the top figure corresponds to mill operation at the lowest energetic conditions, the middle figure to the medium energetic conditions, and the bottom figure to the highest energetic condition. It is clear that the temperature profiles changed drastically at different energetic (Pw: power consumption) levels. While the lowest Pw caused a 6 oC increase at the end of 60 min milling, the highest Pw caused a 28 oC jump in the temperature in around 2 min. After evaluating all specific times and specific temperatures (not shown for brevity here), the time required for the temperature to reach 22 oC and the temperature at 5 min. of milling were found to be the best measures to represent all dynamic data and a strong correlation between these measures and Pw was developed with R2~0.9. This correlation is important because if one can develop a correlation between the power number and the process conditions, one can make power predictions and then specific temperature predictions. Our experimental results have also pointed out that an increase in stirrer speed and/or bead loading led to significantly notable max. temperature and faster temperature rise, whereas the impact of bead size was relatively weak within the range studied. Overall, this study has provided the first thermal analysis of the WSMM process for nanomilling drug suspensions, demonstrating the criticality of process parameter selection to avoid any overheating or multiple mill shutdowns. We have also demonstrated the fundamental origin of reduced/restricted design space for drug suspension manufacturing due to limits imposed by the thermal effects due to high heat generation at the higher speeds/bead concentrations.

References

[1] A. Bhakay, M. Rahman, R.N. Dave, E. Bilgili, Bioavailability enhancement of poorly water-soluble drugs via nanocomposites: Formulation–processing aspects and challenges, Pharmaceutics, 10 (2018) 86.

[2] M. Li, M. Azad, R. Davé, E. Bilgili, Nanomilling of drugs for bioavailability enhancement: A holistic formulation-process perspective, Pharmaceutics, 2016.

[3] L. Peltonen, Design space and qbd approach for production of drug nanocrystals by wet media milling techniques, Pharmaceutics, 10 (2018) 104.

[4] E. Bilgili, G. Guner, Mechanistic modeling of wet stirred media milling for production of drug nanosuspensions, AAPS PharmSciTech, 22 (2020) 2.

[5] D. Eskin, O. Zhupanska, R. Hamey, B. Moudgil, B. Scarlett, Microhydrodynamic analysis of nanogrinding in stirred media mills, AIChE Journal, 51 (2005) 1346-1358.

[6] N. Parker, M. Rahman, E. Bilgili, Impact of media material and process parameters on breakage kinetics–energy consumption during wet media milling of drugs, European Journal of Pharmaceutics and Biopharmaceutics, 153 (2020) 52-67.

[7] G. Guner, D. Yilmaz, E. Bilgili, Kinetic and microhydrodynamic modeling of fenofibrate nanosuspension production in a wet stirred media mill, Pharmaceutics, 13 (2021) 1055.

[8] S. Verma, S. Kumar, R. Gokhale, D.J. Burgess, Physical stability of nanosuspensions: Investigation of the role of stabilizers on ostwald ripening, International Journal of Pharmaceutics, 406 (2011) 145-152.

[9] B. Van Eerdenbrugh, L. Froyen, J. Van Humbeeck, J.A. Martens, P. Augustijns, G. Van den Mooter, Drying of crystalline drug nanosuspensions—the importance of surface hydrophobicity on dissolution behavior upon redispersion, Eur. J. Pharm. Sci., 35 (2008) 127-135.

[10] Nisso- hydroxypropyl cellulose. Technical data sheet 2011.

[11] C. Konnerth, C. Damm, J. Schmidt, W. Peukert, Mechanical activation of trans-stilbene during wet grinding, Advanced Powder Technology, 25 (2014) 1808-1816.

[12] C. Knieke, M.A. Azad, R.N. Davé, E. Bilgili, A study of the physical stability of wet media-milled fenofibrate suspensions using dynamic equilibrium curves, Chemical Engineering Research and Design, 91 (2013) 1245-1258.

[13] G. Guner, M. Kannan, M. Berrios, E. Bilgili, Use of bead mixtures as a novel process optimization approach to nanomilling of drug suspensions, Pharmaceutical Research, 38 (2021) 1279-1296.

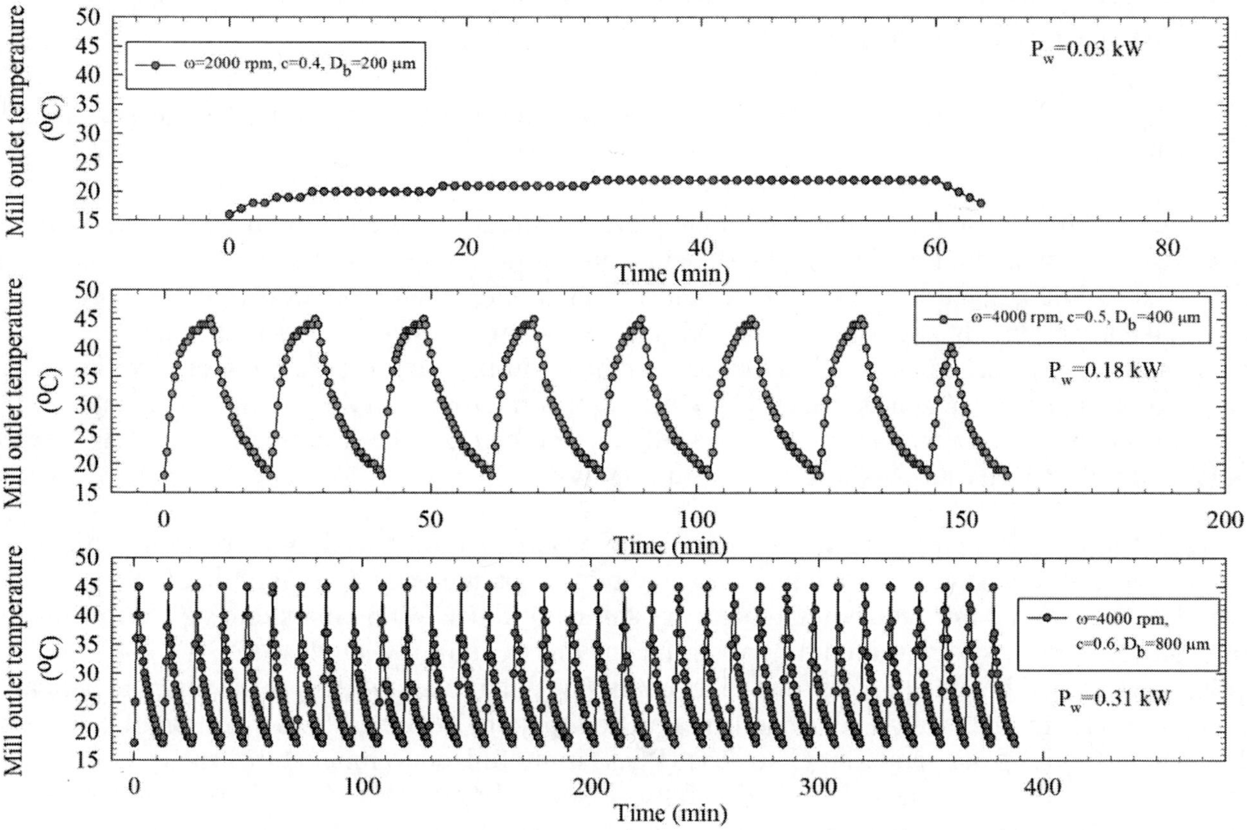

Mill outlet temperature (°C)
ω=2000 rpm, c=0.4, Db=200 μm
Pw=0.03 kW
Time (min)
ω=4000 rpm, c=0.5, Db=400 μm
Pw=0.18 kW
Time (min)
ω=4000 rpm, c=0.6, Db=800 μm
Pw=0.31 kW
Time (min)

139e Magnetophoretic Solid-Solid Separation of β-Lactam Antibiotics from Biocatalyst on a Pilot Plant Scale

Colton Lagerman, Grant Marshall, Matthew McDonald, Martha Grover, Ronald Rousseau, Andreas Bommarius

Enzymatic reactive crystallization is a promising green synthesis route for the production of semi-synthetic β-lactam antibiotics. Enzymatic *β*-lactam antibiotic synthesis is also a focus for continuous pharmaceutical manufacturing as a study to improve pharmaceutical production quality through continuous processing. In addition, simulations have shown that use of continuous reactive crystallization for β-lactam antibiotic synthesis can be optimized to increase productivity and conversion compared to batch synthesis [1]. One major limitation is in the use of a solid catalyst (immobilized enzyme) in the presence of crystallizing product and the resulting solid-solid separation that is required to purify the API and recycle the catalyst.

Solid-solid-liquid separations are uncommon design problems rarely used in the pharmaceutical industry. Magnetic separation is one possible method to separate enzyme immobilized to paramagnetic material from the nonmagnetic crystalline product. While magnetic separation has been implemented in prior systems, including with the enzyme penicillin G acylase (PGA) [2-4], the application detailed herein is unique due to its continuous nature and its use at the pharmaceutical pilot plant scale. In this work, a continuous magnetic separator designed for separation of synthesized β-lactam antibiotics from penicillin G acylase (PGA) immobilized to magnetic particles is demonstrated on a pilot plant scale. The separators operate through use of neodymium magnets to generate a stable magnetic field gradient for magnetophoresis, allowing for isolation of crystallizing product from magnetic particles in separate channels. The separators are modular and can be incorporated in series to improve separation efficiency or in parallel to increase throughput.

Separators in series were first demonstrated through separation of paramagnetic particles from amoxicillin crystals generated through a pH-shift crystallization. The magnetic particles enter two separators in series before being recycled back to the crystallizer. The amoxicillin crystals are removed from the system and pass through a magnetic trap to capture any remaining magnetic particles before being collected and analyzed for purity. PGA was then immobilized to functionalized magnetic particles for demonstration of the reactive crystallization of amoxicillin and subsequent separation of amoxicillin trihydrate from magnetic catalyst. Amoxicillin is synthesized through coupling of a β-lactam ring donor, 6-aminopenicillanic acid (6-APA) with an activated acyl donor, hydroxyphenylglycine methyl ester (HPGME) catalyzed by the immobilized penicillin G acylase. In this system, amoxicillin is simultaneously crystallized out of solution. The slurry of immobilized PGA and amoxicillin are then fed to a series of separators where pure amoxicillin crystals are isolated for washing and drying and the magnetic catalyst particles are recycled back to the reactor for reuse as demonstrated in the pH shift crystallization.

Amoxicillin trihydrate was generated and isolated continuously in a mixed suspension–mixed product removal (MSMPR) at the pilot plant scale for 10 hours. The separators were found to be capable of processing >700 mL/hour of amoxicillin and catalyst slurry with a separation efficiency of >99.9% upon incorporation of a magnetic trap in both the pH-shift crystallization and reactive crystallization. Final amoxicillin trihydrate was analyzed for iron oxide content to determine purity. While the system was demonstrated for amoxicillin trihydrate, magnetic separation could be incorporated for synthesis of other β-lactam antibiotics by PGA and likely be tuned for other continuous or batch systems where multiple solid phases are present. Scaling the separation is also possible through use of separators in parallel and will be a focus in future studies.

1. McDonald, M.A., et al., *Continuous reactive crystallization of β-lactam antibiotics catalyzed by penicillin G acylase. Part I: Model development.* Computers andamp; Chemical Engineering, 2019. **123**: p. 331-343.
2. Luo, X. and L. Zhang, *Immobilization of Penicillin G Acylase in Epoxy-Activated Magnetic Cellulose Microspheres for Improvement of Biocatalytic Stability and Activities.* Biomacromolecules, 2010. **11**(11): p. 2896-2903.
3. Netto, C.G.C.M., H.E. Toma, and L.H. Andrade, *Superparamagnetic nanoparticles as versatile carriers and supporting materials for enzymes.* Journal of Molecular Catalysis B: Enzymatic, 2013. **85-86**: p. 71-92.
4. Liu, R., et al., *Covalent immobilization and characterization of penicillin G acylase on magnetic Fe2O3/Fe3O4 heterostructure nanoparticles prepared via a novel solution combustion and gel calcination process.* International Journal of Biological Macromolecules, 2020. **162**: p. 1587-1596.

139f Innovative Manufacturing Strategy Enabling Filament-Based 3D-Printing of Lipid-Based Advanced Dosage Forms

Moaaz Abdelhamid, MSc, Carolina Corzo, Martin Spörk, Ioannis Koutsamanis, Carolina Alva, Ana Belén Ocampo, Mira Maisriemler, Eyke Slama, Dirk Lochmann, Sebastian Reyer, Tanja Freichel, Sharareh Salar-Behzadi

Motivation and Scope

In the last decade, additive manufacturing, commonly known as 3D printing, has captured significant attention in the pharmaceutical industry for its ability to transform the production pipeline. The technological advancement offered by 3D-printing made the transition from large-scale production toward a more personalized model possible. Therefore, 3D-printing has the potential to fulfill the requirements of personalized therapies by enabling on-demand adjustable dosage form production that cannot be afforded using traditional large-scale manufacturing technologies such as tableting. However, many complications can impact and limit the applications of 3D printing in the pharmaceutical sector. For example, the relatively limited number of excipients, mainly polymer-based, that are employed for pharmaceutical 3D-printing diminish its capabilities. Therefore, extending the portfolio of excipients with advanced functionality is necessary. For certain 3D-printing technologies, such as fused deposition modeling (FDM^{TM}), the lack of printing-associated properties of the intermediate product (filament) is another challenging complication. Hence, developing more effective and advanced technologies to produce printable filaments is required. This study demonstrates the optimization of the extrusion step to develop high-quality drug-loaded lipid-based filaments with desired printing properties and provide accurate and more efficient active pharmaceutical ingredient (API) loading in the lipid matrix.

Materials

Polyglycerol ester of fatty acids, specifically PG6-C16 partial ester (PG6C16p), composed of six glygerol moieties partially esterified with palmitic acid and commercially available as Witepsol® PMF166 was selected as the lipid-based excipient for the embedding of API.

Methods:

Extrusion: A lab-scale twin-screw extruder equipped with two co-rotating screws and three heating barrels was used to produce the filaments. A conveyor belt was used to provide a controlled movement of the produced filament and to adjust its diameter. Besides, extrusion process parameters such as screw speed and nozzle diameter were adjusted to reach the targeted filament dimensions suitable for 3D-printing. Two different feeding strategies during extrusion were screened and compared with each other:

A. Separate feeding of molten lipid and solid API

PG6C16p was fed in its molten state using a liquid pump, while API was fed using the standard powder feeder. The feeding rates of both the liquid pump and the powder feeder were adapted to reach the desired loading inside the filament. In this approach, both materials meet at the extruder inlet. The mixing step and extrusion happen inside the extruder barrels (semi-cold extrusion).

B. Feeding a single molten mixture of lipid and API

The second feeding approach was feeding both materials together as a single molten mixture. A mixture of PG6C16p and API (8% w/w) was prepared by weighing, mixing, and melting. The molten mixture was solidified and used afterwards as the feeding mixture for extrusion. The feeding mixture was again molten and fed to the extruder at 90 °C using the liquid pump. The extrusion temperature was set at the onset of crystallization (cold extrusion).

3D-Printing: A conventional, low-cost filament-based 3D-printer was customized and used for the 3D-printing of lipid-based API-loaded tablets from the manufactured filaments.

Characterization of filaments and 3D-printed dosage forms: The flexibility of filaments was screened by measuring their critical bending diameter. The solid-state of the lipid and the API as well as the alteration in the crystallinity of the API were assessed in the filament and the 3D-printed tablets via DSC and SWAXS.

Results

Filament manufacturing using the first strategy in which the excipient and the API are fed separately was feasible, yet impractical for many reasons. The fine API powder showed poor flowability and hence inaccurate dose loading in the filaments. Mixing of the API powder with molten lipid before reaching the extruder inlet resulted in bridging and interrupted process. The produced filaments displayed peeling resulting in a rough surface and, therefore, unprintable due to the lack of transportability via the driving wheels of the 3D-printer. On the other hand, the second strategy was very efficient. Only one feeding approach (liquid pump) was employed to feed both materials, ensuring a less complex process. The feeding rate was controlled effectively, allowing accurate dosage loading. Neat filaments with smooth surfaces were produced. Melting of API and PG6C16p mixture before feeding to the extruder was also advantageous. For example, the API is amorphized completely before extrusion and dispersed within the lipid matrix. While in amorphous form, the API has a glass transition very close to the onset of melting of the lipid, which allows the mixture to be completely liquid at low temperatures and ready for feeding at 90 °C, below the melting point of the raw API. The glass-forming ability of some APIs becomes effective when fed using this approach, which eventually contributes to a more flexible filament. In this work, the API-loaded filament was found at least 30 times more flexible than the blank lipid filament when produced using the later approach. Furthermore, API peaks in SWAXS and endothermic peaks in DSC were not detected in the freshly prepared filaments and 3D-printed tablets, confirming that it was completely amorphous in the lipid system. After one month, the API remains amorphous in the 3D-printed tablets under long-term conditions (25 °C /60% RH). Signs of recrystallization were observed in the tablets stored under accelerated conditions (40 °C /75% RH). The solid-state of the lipid content of the tablet was found stable upon storage with no polymorphic change.

Conclusion

This work demonstrates effective manufacturing and 3D-printing of drug-loaded lipid-based filaments. It is an essential step toward extending the portfolio of 3D-printing excipients. The advanced extrusion approach opens the door to overcome problems such as poor powder flowability without the need of glidants or the lack of mechanical properties in some approved pharmaceutical polymers without relying on plasticizers. This work is a strong step forward toward the personalization of medicine.

Acknowledgments

This work was funded through the COMET programme which is operated by the Austrian Research

Promotion Agency (FFG) on behalf of the Federal Ministry for Transport, Innovation and Technology (BMVIT) and the Federal Ministry for Digital and Economic Affairs (BMDW). It is also funded by Land Steiermark and the Styrian Business Development Agency (SFG).

139g Ethylene Vinyl Acetate – How to 3D-Print Personalized Implantable Drug Delivery Systems

Simone Eder, Lisa Kuchler, Eva Roblegg, Matthias Katschnig, Martin Spörk

Implantable and insertable drug delivery systems (IDDS) frequently use ethylene vinyl acetate (EVA) co-polymers as carriers due to i) their broad applicability, ii) their potential for long-time controlled drug delivery (i.e., up to several years), and iii) their favorable inflammatory characteristics [1]. Intra-vaginal IDDS use EVA with a comparatively high vinyl acetate (VA) content of 28 % not only to control the drug release but also to provide suitable mechanical characteristics [2], which is of utmost importance for intra-vaginal IDDS [3]. These IDDS are manufactured via extrusion technologies, which limit the IDDS structure to rather simple designs and lacks the potential of customization. The authors recently showed that filament-based 3D-printing serves the purpose of manufacturing customized vaginal inserts of complex geometry – so called vaginal pessaries used to treat stress urinary incontinence via mechanical stabilization – based on a thermoplastic co-polyester [4,5] and different grades of thermoplastic polyurethanes [6]. Similarly, EVA was reported to be processable via 3D-printing, which was however limited to i) EVAs with a VA content lower than 18 % and ii) rather simple designs that do not account for any customization potential [7]. The present study aims at tackling these shortcomings by the establishment of an advanced 3D-printing process together with a custom-tailored hardware to enable the fabrication of complex, customized structures made from EVA with a high VA content. In a first set or experiments, EVA 28 was processed into filaments via hot melt extrusion that were subsequently 3D-printed into urethra pessaries using a 3-axis, gantry-based 3D-printer equipped with a dual-printhead tailored for highly elastic materials and able to print two materials in one print job (Hage3D GmbH, Graz, Austria). More precisely, the printhead uses profiled drives (form fit) for fast and reliable filament conveying. Additionally, a low-loss and partly spring-mounted transmission system increases the efficiency of conveyance between the stepper motor and filament drive. Extremely short and friction-free filament paths ensure a kink-free filament feed, even when the filaments are very soft (i.e. Shore A hardness below 80). Finally, the specialized hot-end provides a counterpressure-minimized nozzle zone. During the 3D-printing process set up, the i) nozzle size, ii) printing speed, iii) infill pattern, iv) build platform temperature, and v) cooling cycles were thoroughly evaluated, while keeping the nozzle temperature well below the degradation temperature of EVA (i.e., 230 °C). EVA 28-based urethra pessaries without any surface defects and enclosed air bubbles were manufactured applying i) a minimum nozzle size of 0.6 mm to avoid nozzle clogging, ii) a comparatively low printing speed to avoid plunging of the individual layers, iii) a perimetric infill pattern, and iv) a low build platform temperature (i.e., non-heated printer bed). In a second set of experiments, the combination potential of EVA 28 and EVA 9 (i.e., an EVA with a VA content of 9%) – a stiffer EVA grade that is used in commercial products as thin membrane to control drug release – in one urethra pessary was assessed. In other words, the pessary was partly made from EVA 28 and partly made from EVA 9. Thereby, the pessary mechanics could be tailored predominantly via the EVA 9 part; more specifically, via i) its fraction, ii) its location in the pessary and iii) its infill design. In a third set of experiments, the part made of EVA 28 was replaced by EVA 28 loaded with a model drug. Thereby, the drug release properties could not only be modified via the EVA 28/drug part, including its fraction and location in the pessary but also via the spatial arrangement of the EVA28/drug and the EVA 9 part. Summarizing, our results demonstrate that 3D-printing is a promising technology to develop customized, complex IDDS using the well-established pharmaceutical polymer EVA. Customized hardware design together with proper process design and innovative multi-material concepts facilitated individualization of both, the mechanical properties and the release patterns of complex pessary structures, which builds the basis for personalized vaginal IDDS products.

References

[1] C. Schneider, R. Langer, D. Loveday, D. Hair, Applications of ethylene vinyl acetate copolymers (EVA) in drug delivery systems, J. Control. Release. 262 (2017) 284–295. https://doi.org/10.1016/j.jconrel.2017.08.004.

[2] F.J.M.E. Roumen, T.O.M. Dieben, Clinical acceptability of an ethylene-vinyl-acetate nonmedicated vaginal ring, Contraception. 59 (1999) 59–62. https://doi.org/10.1016/S0010-7824(98)00145-0.

[3] P. Boyd, R. Merkatz, B. Variano, R.K. Malcolm, The ins and outs of drug-releasing vaginal rings: A literature review of expulsions and removals, Expert Opin. Drug Deliv. 0 (2020). https://doi.org/10.1080/17425247.2020.1798927.

[4] I. Koutsamanis, A. Paudel, C.P. Alva Zúñiga, L. Wiltschko, M. Spoerk, Novel polyester-based thermoplastic elastomers for 3D-printed long-acting drug delivery applications, J. Control. Release. 335 (2021) 290–305. https://doi.org/10.1016/j.jconrel.2021.05.030.

[5] M. Spoerk, F. Arbeiter, I. Koutsamanis, H. Cajner, M. Katschnig, S. Eder, Personalised urethra pessaries prepared by material extrusion-based additive manufacturing, Int. J. Pharm. 608 (2021). https://doi.org/10.1016/j.ijpharm.2021.121112.

[6] S. Eder, L. Wiltschko, I. Koutsamanis, J.A.A. Urich, F. Arbeiter, E. Roblegg, M. Spoerk, Toward a new generation of vaginal pessaries via 3D-printing: concomitant mechanical support and drug delivery, Euopean J. Pharm. Biopharm. (2022).

[7] N. Genina, J. Holländer, H. Jukarainen, E. Mäkilä, J. Salonen, N. Sandler, Ethylene vinyl acetate (EVA) as a new drug carrier for 3D printed medical drug delivery devices, Eur. J. Pharm. Sci. 90 (2016) 53–63. https://doi.org/10.1016/j.ejps.2015.11.005.

208a *In silico* Design of a Commercial-Scale Bioreactor for Optimum Oxygen Transport Using Computational Fluid Dynamics (CFD)

Iman Mirzaee, Mark Brothers, Gregory Stimpfl, Saeed Jafari Kang, Maryam Medghalchi, Fabrice Schlegel, Pablo A. Rolandi

In this work, a holistic *in silico* bioreactor model is employed to optimize oxygen mass transfer and homogeneity in a novel commercial scale stirred tank bioreactor using multiphase Computational Fluid Dynamics coupled with a Population Balance Model (CFD–PBM). The bubble population is represented by the Method of Moments (MOM) to account for bubble size distribution including breakup and coalescence of the bubbles. Hydrodynamics and species transport equations are solved to obtain oxygen mass transfer and its transport through the simulated cell culture media, and it includes a cellular oxygen uptake rate derived from a metabolic model of cell culture. The mass transfer coefficient is obtained using a combination of literature turbulent based theories and comparisons against experimental data for mass transfer (kla) measurements based on dissolved oxygen (DO) in the commercial scale bioreactor. The model is validated by comparing the numerically predicted kla values and the experimental data. The validated model is then applied to investigate kla and DO heterogeneities in the commercial bioreactor, in addition to oxygen transport performance optimal design. The model is used to iterate the underlying design more than 20 times, by running in excess of 200 simulations in a matter of weeks.

208b Automated Reaction Optimization Under Dynamic Flow Conditions

Brian Wyvratt, Jonathan P. McMullen

Automated optimization in flow reactors is a technology that continues to gain interest in academic and industrial research. For drug substance applications, where limited material is available for extensive studies, it is imperative that the automated optimization procedure identify ideal conditions for manufacturing in a resource sparing manner. It is equally as important that these investigations provide data-rich results so that the information can be used for process understanding. Achieving these two objectives in parallel is challenging with traditional automated optimization systems that rely on steady-state data. Dynamic flow systems, which adjust reaction inputs in a controlled manner to collect transient reaction results, maximize reaction information content. In this work, the benefits of automated optimization flow reactors and dynamic flow systems are combined to demonstrate the synergistic gains in knowledge collection and reaction information using a nucleophilic aromatic substitution as a case study. Using a gradient-based search algorithm, the reaction is optimized using a multi-faceted objective function that accounts for yield, material input, and productivity. Results from the automated dynamic optimization were used to establish a reaction model to provide great insight in the reaction kinetics and selectivity.

208c Self-Supervised Learning Methods for Drug Substance and Drug Product Characterization in the Pharmaceutical Industry

Hossein Salami, Daniel Skomski

Machine learning methods have been applied to a variety of problems in chemical engineering research and development. Among these are data-driven, neural network-based methods that are the standards for most computer vision related tasks. Typically in the form of convolutional networks, these are useful tools for various tasks such as classification and segmentation. In the context of the pharmaceutical industry, such tasks include analyzing raw image data generated by testing samples of drug product and drug substance for different modalities. Examples include microfluidic or powder-dispersed optical imaging data (for characterizing particles in sterile liquid formulations and oral formulations) as well as data generated from in situ cameras in crystallization vessels. Naturally, such characterizations have important implications regarding the regulatory aspects of product development.

Data-driven in nature, these models usually rely on large amounts of data to achieve goals such as classifying subvisible particles in a solution or detecting extraneous matter or impurity crystals in a vessel. However, training these models for such tasks requires labeled data that needs to be prepared by a human user, which can be a tedious task and very time consuming. In this talk, we will discuss how one can leverage a family of self-supervised or weakly supervised learning methods to facilitate performing speedy training tasks and thereby accelerate practical applications. These methods include autoencoder-based and contrastive learning-based approaches. In essence, the methods are built on the idea to invoke the networks to perform a pre-text task in which they learn the most important features of the available data without relying on labels provided by an operator. We will discuss applying such approaches to characterizing different systems from protein aggregates in sterile liquid formulations to impurity particles in small-molecule crystallization processes.

208d Development of a High-Throughput Kinetics Protocol and Applications

Xiao Li, Anna Dunn, Robert Yule, Boung Wook Lee

High-throughput experimentation (HTE) has become integral to the pharmaceutical industry, with most major pharmaceutical companies investing in automation and high throughput screening technologies. Testing hundreds of reactions in parallel has distinct advantages; however, one clear disadvantage is that performing a reaction on µM scale is not always indicative of the reaction's performance on multi-kilogram scale. Additionally, a great deal of information is lost by looking at a single time point. Valuable data around intermediates, over-reaction, catalyst induction periods, etc. are invisible to a typical HTE workflow, which involves analyzing reactions at a single time point (e.g. 18 hours). We envisioned a workflow in which time courses for each well of a high-throughput screen were collected. With this change in strategy, it could then become possible to complete high-throughput screening, select reaction conditions, gather kinetic information, and successfully build a kinetic model in less than one week. A kinetic model consisting of scale-independent parameters allows for virtual reaction optimization where the input concentrations, catalyst loading, and temperature can all be simulated and adjusted to understand their impact on yield or quality in a matter of seconds. A case study is presented with a transition metal salts/TMSCl-catalyzed aza-Michael reaction to showcase the performance and robustness of the high-throughput kinetic platform. A Reaction Progress Kinetic Analysis (RPKA) approach is utilized to quickly screen the rates of 48 catalyst/solvent combinations and create a mechanistic model. The first-principles kinetic model provides support for a proposed mechanism of dual activation by TMSCl. This presentation will also showcase some data-driven models (i.e., hybrid-modeling approach) to address some questions that are difficult to be explained by a simple mechanistic model, such as hart-to-detect intermediates and different reaction pathways under different reaction conditions.

208e A Techno-Economic Assessment of Acrylates Downstream Processing with Membrane Extraction

Jie Yang, Anita Buekenhoudt, Miet Van Dael, Patricia Luis, Yamini Satyawali, Robert Malina, Sebastien Lizin

Acrylate esters represent a versatile family of building blocks for thousands of copolymer compositions. This includes butyl acrylate (BA), 2-ethylhexylacrylate (2-EHA), methyl methacrylate (MMA), butyl methacrylate (BMA), and others. Acrylic resins based on these monomers exhibit excellent weather resistance, high gloss and colour retention, and durability. They are primarily used to prepare emulsion and solution polymers. The global market for commodity acrylate esters is over 3 million metric tons, the EU production capacity is currently about 850,000 tons of acrylate esters. 2-Ethylhexylacrylate (2-EHA), is an important bulk chemical used in the production of homo and copolymers and it will be considered as a model acrylate in this work. On industrial scale, 2-EHA is produced by esterification of acrylic acid (AA) and 2-ethylhexanol (2-EH), with formation of water as byproduct. The reaction is performed batch-wise in the presence of an organic solvent and sulfuric acid as catalyst. The plants have capacities in the same order of magnitude of 20,000 t/year of 2-EHA.

For the downstream processing needed for product separation, all remaining base materials and by products resulting from the acrylate synthesis reaction need to be removed. This is typically done in two steps: 1) washing with water to remove the acrylic acid; afterwards, the water-acrylic acid solution is fed into an organic stripper/water distillation column. The left waste from the bottom of the column is either treated biologically or incinerated after concentration; and, 2) the organic solution containing the acrylate product is sent to a distillation unit to remove the product from the solvent. To avoid self-polymerisation and the formation of other by-products, the separations need to be done at a maximum temperature of 50°C. The conventional separation processes are inefficient water-washing and vacuum distillation. However, a large amount of washing water and energy is demanded in this traditional process. An efficient membrane extraction process has potential to replace the current complex and costly distillations, washing and decanting steps. The technical feasibility and advantages of replacing the traditional water wash process with membrane extraction have been demonstrated by Buekenhoudt et al. (2021). But there is not yet a full-scale application for this process. Moreover, the availability of economic assessments of downstream processing with membrane extraction is limited. It is still unclear whether membrane extraction of acrylic acid by is more economically attractive than the traditional waster wash process.

Given that an economic analysis of membrane extraction for acrylate downstream processing is lacking, a techno-economic assessment (TEA) is presented in this paper to understand the main drivers of the manufacturing costs. The water wash process is also analyzed to benchmark the membrane processes against the traditional process. A representative acrylate 2-EHA is taken as the product in the analysis. The TEA consists of four steps: 1) market study, 2) process flow diagram and mass and energy balance, 3) economic analysis, and 4) uncertainty analysis. Step 1 identifies the market trends, related prices, and competitive processes, etc. Step 2 describes the alternatives in more detail based on their process flow diagrams. It also outlines the modelling assumptions (a) that are key to enable comparability across the evaluation of the selected alternatives and (b) that are specific to each alternative. Further, in Step 2, the mass and energy balances of the two processes are calculated by identifying and quantifying the different input and output streams. Step 3 assesses the economic profitability of the two cases by estimating the manufacturing cost. The technological and economic parameters are integrated by linking the prices of the inputs and outputs to the mass and energy balance of the process. Step 4 entails a Monte Carlo sensitivity analysis to identify the impact of the key parameters on the manufacturing cost. The influence of the input parameters is evaluated by drawing 10,000 observations from the respective distributions and recalculating the manufacturing cost, using the Crystal Ball extension. A triangular distribution (−10%; +10% from the default values) was applied to the input parameters. The rank order correlation can then be used to identify which of the input parameters causes the most variation in the manufacturing cost. This is followed by a local sensitivity analyses on those parameters.

A solute AA distribution equation was formulated to determine the amount of extractant water needed, which determines the cost of the AA removal step in the two processes. The results, shown in Figure 1, indicate that the two processes have the same cost of €1.70/kg 2-EHA for the esterification step. The calculated downstream processing costs of the two processes were €0.87/kg 2-EHA for the membrane process and €0.60/kg 2-EHA for the water wash process. The personnel cost was found to have the largest share in the cost for both processes, which is €0.28/kg 2-EHA. For the water wash process, the cost for AA removal (€0.21/kg 2-EHA) has the second largest contribution. This indicates that there is a margin for improvement from the economic point of view in the water wash step for the traditional downstream processing. Correspondingly, the cost for AA removal in the membrane process has a much higher CapEx for the membrane (€0.25/kg 2-EHA) than the waster wash column (€0.02/kg 2-EHA). This is mostly related to the membrane cost (€1,500/m2) and the membrane area needed (336 m2). However, the membrane process has a lower cost for energy (€0.08/kg 2-EHA) and waste disposal (€0.09/kg 2-EHA) than the water wash process. This is because less extractant water is needed for the membrane extraction process. Therefore, more energy is consumed for the water distillation to concentrate the AA waste solution afterwards, and more AA waste is generated. A global and local sensitivity analysis was carried out to guide the process development by defining the influence of the process parameters on the downstream processing cost. It can be concluded that the membrane parameters have a major impact on the downstream processing cost. This mostly includes the AA flux, cost, extraction time. When the membrane cost is lower than €0.14/kg 2-EHA, the downstream processing cost of the membrane process will be lower than the traditional water wash process.

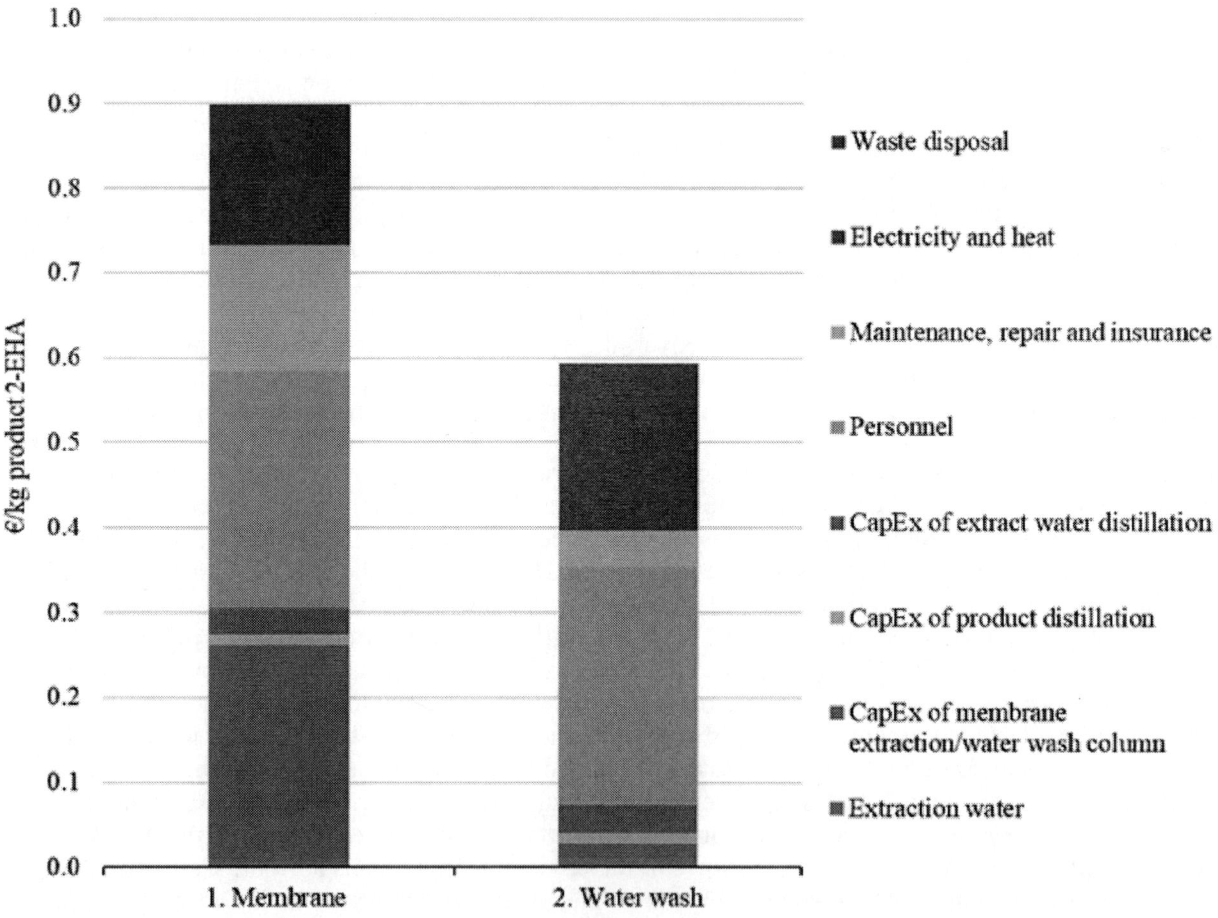

208f Evaluating the Syringeability of Drug Products Using Mechanistic in-Silico Modeling

Saman Seifi, Calvin Clark, Leonela Vega Loaisa, Fabrice Schlegel, Dan Groszmann, Pablo A. Rolandi

The syringeability of drug products refers to determining that the extrusion forces necessary to push the drug product out of the syringe are within acceptable limits. The drug properties as well as means of its delivery are subjected to change during their development cycles and requires repeating characterization testing that takes weeks to complete. Therefore, a mechanistic in-silico model has been developed to compute the extrusion force required for dosing saving tremendous amount of time and cost as a result. The extrusion force is required to overcome two resistive components: friction resistance from the lubricant and drug product resistance from the fill. The friction force depends on the area of contact between the stopper and the syringe, the viscosity of the lubricant, and the film thickness. Finite element contact simulation was used to obtain the area of contact between the stopper and syringe at different pressure levels on the stopper face. Similarly, finite element flow simulation was used to compute the drug pressure on the stopper face. The addition of these two components results in the total extrusion force. This model takes drug properties (density and viscosity), fill volume, syringe and needle geometry, stopper shape and materials, and injection time as inputs. This model can perform large scale DOE (Design of Experiment) type studies by sweeping input parameters across their tolerance band and/or acceptable ranges. This work describes such a high-fidelity in silico model, and present the results of these simulations, including experimental validation. The benefits are arriving at a greater fundamental understanding, which translates into design decisions with greater speed and confidence by product teams.

208g Digital Design and Evaluation of Separation Alternatives for the Green Manufacturing of Lomustine

Daniel Casas Orozco, Daniel Laky, Jaron Mackey, Varun Sundarkumar, Giulia Murbach De Oliveira, David H. Thompson, Gintaras Reklaitis, Zoltan Nagy

Manufacturing of active pharmaceutical ingredients (APIs) typically involves a series of chemical synthesis and separation steps, which together comprise the so-called drug substance (DS) process. For chemical synthesis, batch or continuous flow reactors are used, whereas cooling/antisolvent crystallization is typically employed to recover pure or nearly pure APIs. Often, there are additional intermediate purification/separation steps performed for various purposes. For instance, undesired impurities, precursors or catalysts that may negatively impact subsequent synthesis or purification steps must be removed prior to those affected operations. Also, concentrations of certain toxic species/solvents are required to be reduced to an acceptable level in the final product formulation.

The unit operations involved in these separation/purification steps typically rely on phase equilibrium, with vapor-liquid equilibrium (VLE) - and liquid-liquid equilibrium (LLE) - based equipment most frequently used [1], [2]. These operations have been long studied, and both first-principles and reduced-order models are available for process design and analysis. With the availability of such models and their efficient numerical implementation, systematic in-silico evaluation and comparison of such separation trains can be performed, and a thorough techno-economic analysis of their design and operating conditions via optimization or enumeration techniques can be achieved.

As a way to make unit operation models for pharmaceutical manufacturing readily accessible, we recently developed the Python-based library PharmaPy [3], which has been successfully used for digital design and analysis of a series of pharmaceutical process case studies [4]–[6]. The object-oriented architecture of PharmaPy has made it possible to develop and maintain a diverse set of DS unit operations in different processing modes, viz. batch, semi-batch and continuous operation. For the present work, one-stage flash batch/semi-batch/continuous vaporization and multi-stage, steady state and dynamic distillation models are evaluated. Moreover, single and multiple-stage LLE models are introduced as part of the PharmaPy library, with differing levels of complexity depending on how the coexisting liquid phases are modelled (detailed thermodynamic description or empirical partition coefficients). With the large number of possible unit operation types, operating modes, and operating conditions it is evident that a digital framework for systematic simulation, analysis and optimization of such alternatives is an essential process engineering tool.

In the present work, digital design of the synthesis-solvent switch section for lomustine production is developed using the PharmaPy library. Lomustine is a high-value drug for the treatment of brain cancer, which has seen a significant price increment in the recent years, creating important patient access problems [7]. The most recent efforts of our team have been in developing synthesis routes that lead to efficient continuous manufacturing of lomustine, by replacing traditional solvents by more environmentally-friendly ones in both the chemical synthesis and the separation/purification steps. The case studies presented show that effective lomustine crystallization requires that solvent switch must be performed after the synthesis steps using either VLE or LLE equipment.

In this work, different end-to-end batch, hybrid and continuous flowsheets were systematically enumerated and analyzed in terms of process economy and green chemistry metrics. To this end, a logical system first generates all the possible reactor – separation technology combinations between continuous plug-flow reactors (PFRs) or batch reactors and their corresponding VLE or LLE batch or continuous separation equipment. Then, PharmaPy flowsheet objects of each alternative process pathway are created and their design and operating conditions determined via a simulation/optimization framework. Significant improvements in terms of process feasibility for downstream processing are achieved when continuous staged distillation was used instead of flash separation, where the inherent thermodynamic limitations of one-stage VLE separations are overcome. LLE also proves to be an attractive option to VLE separation given the negligible lomustine solubility in the extraction solvent used. In terms of economics, continuous processing exhibits the lowest capital investment but at the cost of losses of process materials during start-up times. Thus, campaign length again is shown to be an important decision variable in the digital design.

References

[1] K. P. Cole *et al.*, "Kilogram-scale prexasertib monolactate monohydrate synthesis under continuous-flow CGMP conditions," *Science (80-.).*, vol. 356, no. 6343, pp. 1144–1150, Jun. 2017.

[2] S. Diab, N. Mytis, A. G. Boudouvis, and D. I. Gerogiorgis, "Process modelling, design and technoeconomic Liquid–Liquid Extraction (LLE) optimisation for comparative evaluation of batch vs. continuous pharmaceutical manufacturing of atropine," *Comput. Chem. Eng.*, vol. 124, pp. 28–42, 2019.

[3] D. Casas-Orozco *et al.*, "PharmaPy: An object-oriented tool for the development of hybrid pharmaceutical flowsheets," *Comput. Chem. Eng.*, vol. 153, p. 107408, Oct. 2021.

[4] D. Casas-Orozco *et al.*, "Digital Design of a Lomustine Manufacturing Process Using PharmaPy," *2021 AIChE Annual Meeting*. Boston, p. 2, 2021. Paper 317f

[5] D. J. Laky, D. Casas-Orozco, C. D. Laird, G. V. Reklaitis, V. Wang, and Z. K. Nagy, "Determination of probabilistic design spaces in the hybrid manufacture of an active pharmaceutical ingredient using the Python-based framework PharmaPy," in *14th International Symposium on Process Systems Engineering*, 2022, p. 2.

[6] I. Hur, D. Casas-Orozco, and Z. K. Nagy, "Dynamic Flowsheet Simulation and Application of Soft Sensors on an Intensified and Integrated Purification Step for Pharmaceutical Upstream Manufacturing," *2021 AIChE Annual Meeting*. Boston, 2021. Paper 117f

[7] Z. Jaman, T. J. P. Sobreira, A. Mufti, C. R. Ferreira, R. G. Cooks, and D. H. Thompson, "Rapid On-Demand Synthesis of Lomustine under Continuous Flow Conditions," *Org. Process Res. Dev.*, vol. 23, no. 3, pp. 334–341, 2019.

566a Manufacturing of Aglycosylated Monoclonal Antibodies in Yeast

Neil C. Dalvie, Joseph R. Brady, Christopher Naranjo, Sergio Rodriguez Aponte, Ryan Johnston, J. Christopher Love

The COVID-19 pandemic revealed insufficient global manufacturing capacity for biologic medicines including vaccines, antiviral drugs, and monoclonal antibodies (mAbs). Before the pandemic, mAbs were the fastest growing class of biologic medicines by number of new approved products and biosimilars, number of patients, and total revenue. This rapid growth is in part due to platform manufacturing processes based on expression in Chinese hamster ovary (CHO) cells and purification with Protein A chromatography. While reliable, CHO cells require complex culture media, stringent sterilization, and viral filtration of the secreted mAb product, all of which contribute to high manufacturing costs. Additionally, development of a stable CHO cell line for manufacturing may take six months or more, due to the slow growth rate. Indeed, development of new mAb treatments for COVID-19 has been outpaced by evolving variants of SARS-CoV-2. New platforms for manufacturing mAbs with shorter development timelines, lower production costs, and larger scales are needed to increase global access and improve pandemic response.

Single-celled eukaryotic hosts such as yeast may be promising alternatives to CHO cells for mAb manufacturing. Yeasts grow quickly to high cell densities on inexpensive media, which enables low-cost fermentation processes. The yeast *Komagataella phaffii* (*Pichia pastoris*) is particularly suited for mAb manufacturing because it contains an advanced secretory pathway, secretes few native host-cell proteins, and is routinely used to manufacture therapeutic proteins like insulin and subunit vaccines. In 2020, the FDA approved eptinezumab (Vyepti), an aglycosylated mAb manufactured by Alder/Lundbeck in *K. phaffii* for treatment of migraine. Aglycosylated mAbs are a growing product class, and could be developed, in theory, for any indication that does not require a particular effector function, such as antiviral therapy. Further engineering and characterization are needed to encourage the widespread adoption of *K. phaffii* and other microbial hosts over traditional CHO cell processes.

Here, we report a platform for reliable manufacturing of aglycosylated mAbs in *K. phaffii*. We developed a modular vector for simple integration and expression of both protein chains, and identified an effective signal peptide for secretion. We assessed the quality of mAbs secreted from *K. phaffii*, and engineered a human IgG1 mAb backbone sequence with minimal modifications that eliminate most product-related variants. We integrated five different aglycosylated mAbs into a strain of *K. phaffii* that we previously engineered to have enhanced secreted productivity, and demonstrated consistent production of high-quality drug substance at lab scale. Finally, we demonstrated fed batch and perfusion manufacturing of mAbs using both *K. phaffii* and CHO platforms. We observed comparable product quality between the two hosts. While the total titer was higher for CHO-based processes, the space-time yield (g/L/day) was comparable between CHO and the microbial host.

These results demonstrate a reliable platform for expression of multiple mAbs in yeast. Microbial hosts, generally, will enable manufacturing of mAbs and other therapeutic proteins at low costs and large scales that are not achievable with mammalian expression systems. The speed of lab scale expression and cell line development, additionally, will enable rapid development and preclinical testing of new antibody products to reduce development costs and assist with pandemic response.

566h Online/At-line Optical Density-based Analytical Method to Quantify Empty-Full Adeno-associated Virus Particles

Srivatsan Ramesh, Ashton Lavoie, Amod Joshi, Jacob Irwin, Eric J. Yearley, Michael Guerrero

The percentage of full adeno-associated virus (AAV) vector particles is quantified using the gold standard analytical ultracentrifugation (AUC) method. However, alternative techniques using (i) the capsid titer (using capsid ELSIA) and vector genome titer (using ddPCR/Q-PCR), (ii) transmission electron microscopy, and (iii) electrophoresis have been employed for % full determination [1]. These techniques can be expensive, laborious, reagent intensive, time-consuming, and not amenable to in-line or at-line monitoring. Owing to the wide variability of samples in terms of titer, buffer matrix, and % full, it is crucial to develop a quicker and easier characterization technique to quantify the titer and empty/full ratio for *(i)* more rapid, reliable characterization for quick decision making, and *(ii)* act as an orthogonal technique to supplement the existing techniques. In this work, we developed a method for quantifying AAV9 percent full using a simple, direct, and rapid UV-Vis method based on prior art demonstrating a similar approach for AAV2 [2]. The absorbance measurement of the samples at 260 nm (A_{260}), 280 nm (A_{280}), and the A_{260}/A_{280} ratio enable the accurate prediction of % full capsids of process intermediates that were orthogonally verified by AUC and ddPCR / capsid ELISA. Highly pure full and empty AAV9 preparations were used to generate this serotype-specific model. These were produced using a hybrid sample preparation method, comprising ultracentrifugation and automated, UV-based fractionation. The developed model enables rapid, online/at-line, and accurate estimation of the fraction of full AAV9 vector particles using UV-Vis absorbance, circumventing time-consuming and manually intensive offline assays.

References

[1] Xiaotong Fu, Wei-Chiang Chen, Christopher Argento, Peter Clarner, Vinay Bhatt, Ryan Dickerson, George Bou-Assaf, Meisam Bakhshayeshi, Xiaohui Lu, Svetlana Bergelson, and John Pieracci. Analytical Strategies for Quantification of Adeno-Associated Virus Empty Capsids to Support Process Development. *Human Gene Therapy Methods*, 2019, 30, 4, 144.

[2] Jurg Sommer, Peter Smith, Sumathy Parthasarathy, Jesse Isaacs, Sharmila Vijay, Jane Kieran, Sharon Powell, Alan McClelland, and Fraser Wright. Quantification of adeno-associated virus particles and empty capsids by optical density measurement. *Molecular Therapy*, 2003, 7, 1, 122.

566e Tuning the Biophysical Properties of an IgG1-Based Antibody-Drug Conjugate By Traversing the Formulation Design Space

Sasha B. Ebrahimi, Xuan Hong, James Ludlow, Dany Doucet, Renuka Thirumangalathu

Antibody-drug conjugates (ADCs) have recently emerged as a powerful class of chemotherapeutics. ADCs consist of an antibody covalently linked to a drug payload, most commonly a small molecule. These structures combine the ability of antibodies to enter cells via selective binding to cell surface receptors with the ability of conjugated drug molecules to elicit cytotoxicity, yielding highly effective therapeutic agents that can kill target cells. Importantly, ADCs have enhanced therapeutic potency and targeting ability compared to either the antibody or drug payload alone. In this regard, as of January 2022, 11 ADCs have been granted approval by the FDA for a range of clinical indications.

While ADCs offer a promising approach for treating a wide variety of diseases, they also present challenges related to structural stability. In particular, the attachment of payload to antibodies is generally achieved via modification of chemically reactive residues. One common strategy for this conjugation consists of partially reducing the antibody's interchain disulfide bonds to yield reactive free cysteines that can react with the linker/drug moiety. However, interchain disulfide bonds play an important role in forming the quaternary structure of antibodies, and therefore reducing these bonds can negatively impact the overall stability of the structure. Moreover, in the case of small molecule payloads, their attachment may introduce an additional hydrophobic moiety that can promote increased self-interaction and aggregation.

This talk will explore our recent work studying an ADC composed of an IgG1 antibody conjugated to an auristatin-based small molecule. The small molecule was attached at cysteine residues of the antibody to yield conjugates with an average drug loading of four as quantified by hydrophobic interaction chromatography. This talk will focus on and discuss the differential behavior observed of the ADC compared to the naked (unconjugated) mAb towards gaining fundamental understanding of how conjugation of surface payload drives structural changes that manifest as differences in product attributes.

One important property of antibody formulations is their degree of opalescence (turbidity) in solution. High opalescence can signal the presence of reversible or irreversible aggregates that impact therapeutic efficacy or elicit *in vivo* toxicity. As a preliminary experiment, the naked mAb and ADC were formulated in citrate buffer to assess their turbidity. In contrast to the naked mAb which exhibited opalescence as high as ~20 nephelometric turbidity units (NTU), the ADC exhibited substantially higher opalescence of ~55 NTU when both were formulated at ~55 mg/mL. This observation motivated a systematic study to understand the mechanistic underpinning of this phenomenon. A high-throughput screen across ~100 formulation conditions demonstrated that varying solution properties such as pH, ionic strength, trehalose concentration, and PS80 concentration led to changes in opalescence of the conjugated mAb. Notably, the largest difference in opalescence was observed under conditions with varying pH. In particular, a significant reduction in turbidity of up to 400% was observed as the solution pH was increased from pH 5.2 to pH 7.0. Because the significantly high opalescence was observed at a pH much less than the pI of the ADC (pI ~8.30 - ~9.10 depending on drug loading), this suggests that electrostatic interactions are one of the primary drivers for the turbidity observed. Interestingly, and in contrast to the ADC, the turbidity of the unconjugated mAb remained unchanged across the design space. Moreover, the ADC exhibited higher turbidity than the unconjugated mAb in all conditions analyzed.

It was hypothesized that the observed differences in solution behavior of the ADC versus the unconjugated mAb was a result of their inherent structural differences. Therefore, the extent of changes in the ADC upon small molecule conjugation was next probed by both differential scanning calorimetry (DSC) and intrinsic fluorescence measurements. Melting temperatures generally correlate with protein stability; increasing stability is usually reflected by higher melting temperatures as the amount of energy required to unfold the protein increases. DSC indicated two melting transitions in both the naked mAb and ADC corresponding to (1). the unfolding of the CH2 domain of the Fc fragment and (2). the unfolding of the unconjugated Fab domain and CH3 domain of the Fc fragment. However, in the case of the ADC, the modification of the protein gave rise to a third melting transition corresponding to unfolding of the drug-conjugated Fab domain. In comparison to the unconjugated Fab region, this domain exhibited an approximately 6°C lower melting temperature. Intrinsic fluorescence measurements were next conducted to make high throughput melting temperature readings across formulations that had previously been used for studying turbidity. While the naked mAb had a melting temperature that was largely invariant to the different conditions tested, the ADC melting temperature was heavily dependent on the physical properties of the solution, including a 6°C increase in melting temperature as the formulation pH was varied from 5.2 to 7.0. Importantly, solution conditions that led to decreases in melting temperature were the same as those that led to higher turbidity, suggesting that structural destabilization plays a key role in dictating the opalescence observed in the ADC. The use of static light scattering experiments revealed that the ADC also has a substantially lower (up to ~10°C) aggregation temperature compared to the naked mAb. Similarly to melting temperature results, the aggregation temperature decreased in formulations that promoted increased turbidity. Taken together, these studies substantiate that the conjugation of the small molecule moiety leads to structural changes that ultimately give rise to less stable formulations.

An investigation was then done to understand whether these structural differences lead to increased levels of interaction between proteins in solution. It was initially observed that turbidity exhibits a nonlinear dependence on ADC concentration. This is especially interesting considering that normal Rayleigh scattering in dilute media should lead to a linear dependence between concentration and turbidity, and implies the presence of protein-protein interactions contributing to the behavior of the ADC. To probe this further, measurements to quantify interaction parameters (K_d) were made using dynamic light scattering. These studies showed that the ADC exhibited negative K_d values that were also significantly lower than the naked mAb. This suggests the presence of a greater extent of attractive protein-protein interactions in the case of the ADC compared to the naked mAb. Moreover, varying buffer composition led to significant changes in K_d, wherein those conditions that previously had given rise to more opalescence also exhibited more negative K_d values. These results illustrate that structural differences between the naked mAb and ADC are manifested into increased attractive protein-protein interactions that lead to enhanced opalescence. To gain further insight into these experimental observations, molecular modeling was carried out. These studies suggested specific structural differences between the ADC and the naked mAb towards proposing a hypothesis for increased protein-protein interactions in the ADC. Taken together, it is anticipated that the experimental and simulation results presented here can lead to learnings applicable to other IgG1-based conjugates where drug loading is predominantly at cysteine residues of the Fab region.

566f Mechanistic Insights into the Stabilization of Proteins By Saccharides

Johanna Dieplinger, Christina Moser, Joana Pinto, Amrit Paudel

Background and Motivation:

The drying of proteins might have detrimental impacts on their structure and function. Excipients stabilize proteins during drying mainly by the following mechanisms [1]: water replacement, glass dynamics, and interfacial competition. The first works by the saccharides replacing the hydrogen bonds to the protein. The second leads to reduced protein mobility within the glassy matrices, prohibiting aggregation. The third explains how surface-active excipients keep the protein away from the interface, thereby protecting it from denaturation.

This work aimed to develop a miniaturized screening platform easy to establish in every laboratory that offers fast screening of protein formulations without spending large amounts of material. Likewise, the drying kinetics of one single droplet can be followed in-situ, and enough material to analyze the protein activity can be generated. Moreover, by applying this approach to different protein formulations containing saccharides, it was possible to systematically analyze the impact of these excipients on the droplet evaporation kinetics.

This study shows how saccharides presenting distinct molecular weights affect the drying kinetics of catalase and trypsin, deriving further mechanistic insights on how some excipients successfully stabilize the proteins while others do not.

Materials and Methods:

Materials:

Trehalose dihydrate (TD) (Mw = 378.33 g/mol), Hydroxypropyl β-cyclodextrin, Kleptose® HP ORAL GRADE (HP) (Mw = 1501 g/mol), Hydroxypropyl β-cyclodextrin, Kleptose® HPB ORAL GRADE (HPB) (Mw = 1387 g/mol) and Dextran 40 EP (DEX) (Mw = 40000 g/mol) have been selected based on pre-screening experiments (data not shown). Milli-Q® (TKA Wasseraufbereitungssysteme GmbH) was used as solvent. Catalase from bovine liver (2,000-5,000 units/mg; ~250 kDa) and trypsin from bovine pancreas (≥2,500 units/mg; ~23.8 kDa) were used as model proteins. The concentrations were 0.04 mM for catalase and 0.42 mM for trypsin (10% w/w of enzymes in the formulations).

Miniaturized Screening (MS):

A single droplet drying method was developed based on the sessile drop principle. An EasyDrop equipment (EasyDrop, Krüss GmbH) equipped with Drop Shape Analysis (DSA1 v1.92, Krüss GmbH) was used. A 15µL-droplet was deposited onto a polypropylene film fixed to a Teflon membrane and placed on a hot plate at 75°C ± 2°C. Drying was performed for 15 min, and the resulting pellets were frozen until analysis.

Evaluation of evaporation rates and droplet shape:

As a droplet shape, a spherical cap was assumed, which describes the region of a sphere below or above a flat surface. According to this, the surface (*S*) and the Volume (*V*) were calculated using Eq. (1) [2] and Eq. (2) [2]:

$$S = \pi(a^2 + h^2) \quad \text{Eq. (1)}$$

$$V = (1/6)\pi h\,(3a^2 + h^2) \quad \text{Eq. (2)}$$

Here, *h* describes the height, and *a* represents the radius of the spherical cap. These parameters were obtained manually for each time point, at which pictures were taken with the DSA software during droplet drying.

Using Eq. 3, the aspect ratio (AR) was calculated:

$$AR = h/w \text{ Eq. (3)}$$

Here, h describes the height, and w describes the width of the dried particle. Values closer to 1 indicate a more spherical shape, and values closer to 0 show a needle-like form.

A spherical droplet of the same volume as a spherical cap was assumed. Therefore, using the calculated volume, the radius was obtained:

$$r = (V/4.19)^{1/3} \text{ Eq. (4)}$$

Here, r describes the droplet radius and V the volume.

Eq. (5) describes the function used to obtain the evaporation rate. The droplet diameter (2*r) was plotted against the drying time, and the evaporation rate was obtained as the slope k [3]:

$$d_t^2 = d_0^2 - kt \text{ Eq. (5)}$$

Here, d_t is the droplet diameter at t, d_0 is the initial diameter of the droplet, k represents evaporation rate.

UV/vis spectroscopy:

A Shimadzu UV-Visible Spectrophotometer UV-2700 (Shimadzu GmbH) was used to gain information on the activity of catalase and trypsin after drying them with different saccharides. The spectrophotometer was operated at 25°C. The corresponding buffers were used as blanks. For catalase, decrease in absorbance at 240 nm (ΔA240) was measured for 3 minutes and for trypsin, increase in absorbance at 253 nm (ΔA253) was measured for 70 seconds and plotted against the recorded time (see Fig. 1a and 1b, n=3).

Statistical analysis:

A 1-way analysis of variance (ANOVA) was conducted using the concentrations of different saccharides and their related activities with catalase or trypsin (Microsoft Office Standard 2016). The goal was to find out if there were any differences between the distinct concentrations. A p-value > 0.05 indicates similarities of samples, whereas a p-value < 0.05 indicates that samples are not similar and that the null hypothesis (no difference) is rejected.

Results and Discussion:

Catalase: The dried mass of protein formulations containing HP and HPB did not dissolve completely. It was observed that protein-saccharide formulations containing the same saccharides did not always seem to follow the same trend. Drying of catalase without saccharides did not result in an extended loss of activity. Thus, the addition of saccharides was not critical for protein stabilization. However, some trends could still be observed (see Fig. 1a): For HP, the activity increases with decreasing concentration until 1.5 mM and drops again. For HPB, no clear trend is observed, which indicates the potential difference the degree of substitution might have on how the mechanisms, by the tested cyclodextrins (CDs) might be affected. CDs could either stabilize the protein by substituting it at the interface (i.e., liquid-air, liquid-substrate) or by integrating it in their hydrophobic pocket. These mechanisms might be affected by the degree of substitution of CDs, leading to higher quantities of HPB being needed to stabilize catalase, whereas lower quantities of HP are enough. 1.875 mM DEX seems to stabilize catalase best. This is probably due to the high glass transition temperature of DEX and molecular bonding or steric hindrance driven by viscosity and rigidity. Further research is needed to fully elucidate these mechanisms. TD seems surprisingly variable, and no clear trend could be observed, although TD is well-known to have superior stabilizing effects on proteins. However, in this case, TD only seemed to impact the stability of catalase at 400 mM beneficially. This might be because catalase is a large protein, while TD is a small disaccharide. The evaporation rates increased with increasing concentrations of TD, HP, and HPB, naturally, as the solid content of the solutions increased. The evaporation rate was reduced with an increasing concentration of DEX. We hypothesize that this might be due to DEX binding water. DEX also resulted in more spherically shaped particles with higher volume.

Trypsin: All of the dried droplets dissolved completely. It was observed that protein-saccharide formulations containing the same saccharides behaved quite similarly (see Fig. 1b). Trypsin without saccharides loses activity, and the addition of saccharides seems to be beneficial. Because trypsin is smaller in size, we propose that more protein molecules be present at the interfaces during drying, leading to more degradation and instability compared to catalase. HP, surprisingly, did not have a stabilizing effect on trypsin. Although this has to be further elucidated, the stabilizing effect of CDs with different degrees of substitution is reported to depend on the nature of the protein, and distinct effects can be observed, as observed herein with trypsin and catalase [4]. Compared to catalase, higher DEX concentrations could be used (7.5 mM) to stabilize the smaller enzyme trypsin. Probably, higher concentrations are needed to induce enough steric hindrance. TD stabilized trypsin at 200 mM, a lower concentration than the one required for catalase, again probably because of trypsin's smaller size. The evaporation rate for trypsin formulations was also the highest for TD. The more HP used, the faster, but the more HPB used, the slower the evaporation. Increasing concentration of DEX also slowed down the evaporation rate and resulted in less elongated particles with higher volume.

Implications and outlook:

Trypsin seems more unstable than catalase during drying in absence of stabilizing excipients, possibly due to their different water solubility and size differences influencing the number of molecules present at the interface. This led to different stabilizing effects observed when distinct saccharides were tested, and some insights into their successful use to stabilize proteins with various molecular sizes could be gained. In future, it will be investigated how these observations align with the impact of distinct drying operations like spray drying on the enzyme formulations. Therefore, some relevant protein-saccharide formulations will be selected and spray dried at lab scale. If successful, this could be an essential pre-formulation tool to apply when producing dried biopharmaceuticals.

Literature:

(1) Ziaee, A.; Albadarin, A. B.; Padrela, L.; Femmer, T.; O'Reilly, E.; Walker, G. Spray Drying of Pharmaceuticals and Biopharmaceuticals: Critical Parameters and Experimental Process Optimization Approaches. *Eur. J. Pharm. Sci.* **2019**, *127* (October 2018), 300–318. https://doi.org/10.1016/j.ejps.2018.10.026.

(2) MathWorld, W. Spherical Cap https://mathworld.wolfram.com/SphericalCap.html.

(3) Vehring, R. Pharmaceutical Particle Engineering via Spray Drying. *Pharm. Res.* **2008**, *25* (5), 999–1022. https://doi.org/10.1007/s11095-007-9475-1.

(4) Samra, H. S.; He, F.; Bhambhani, A.; Pipkin, J. D.; Zimmerer, R.; Joshi, S. B.; Middaugh, C. R. The Effects of Substituted Cyclodextrins on the Colloidal and Conformational Stability of Selected Proteins. *J. Pharm. Sci.* **2010**, *99* (6), 2800–2818. https://doi.org/10.1002/jps.22053.

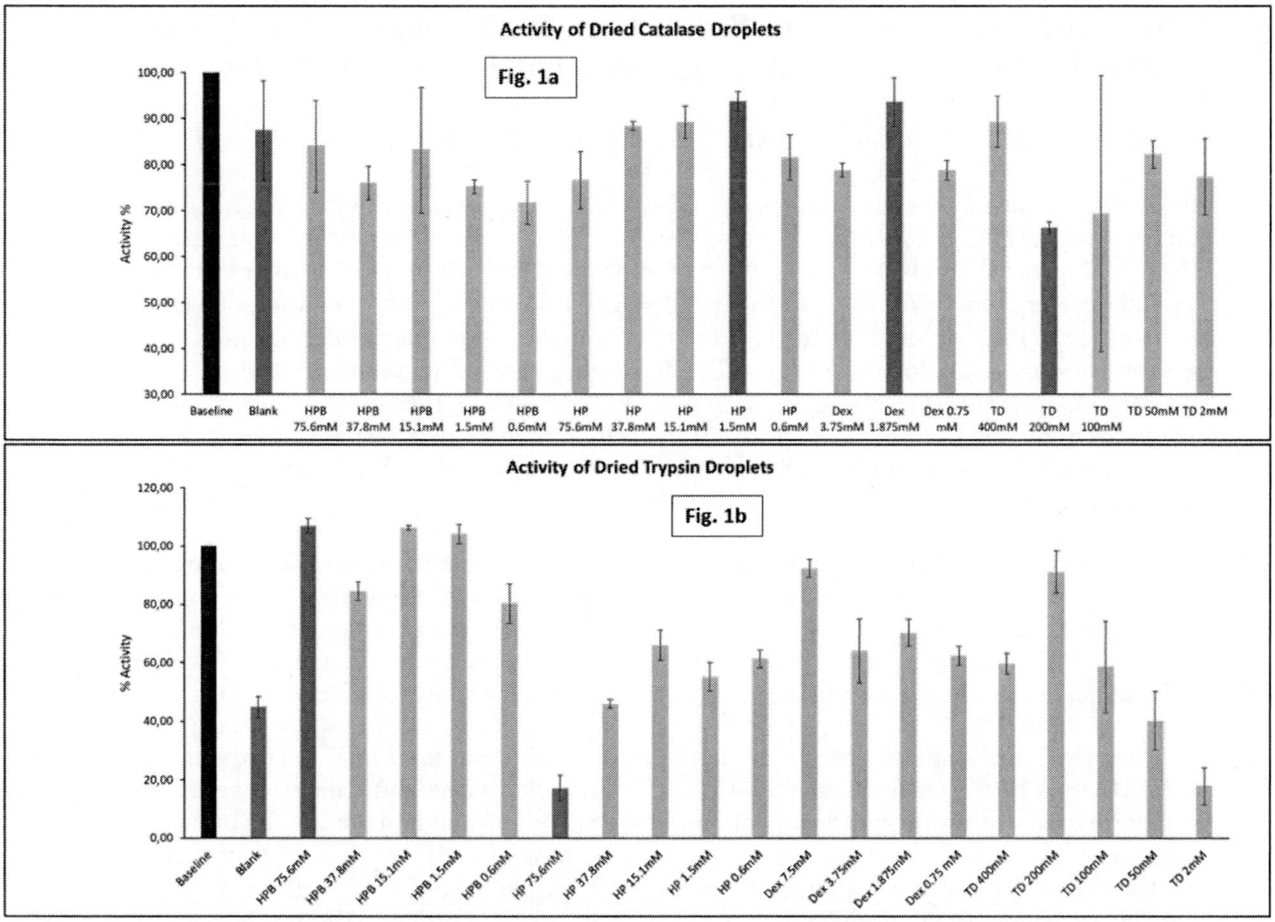

Fig. 1: The activities of catalase (**1a**) and trypsin (**1b**) are presented in dependence on the different saccharides used in variable amounts *(n=3)*. *(Baseline: the raw, non-dried protein material; Blank: the protein dried without the addition of any saccharides;)*

566g A New Approach Towards the Production of a Personalizable Dry Solid Dosage Form for Biologics: In-Vial Direct Dosing and Drying By Inkjet Printing

Daniela Fiedler, Carolina Alva, Joana T. Pinto, Martin Spörk, Ramona Jeitler, Eva Roblegg

The interest in therapeutical relevant proteins and peptides is increasing steadily, not only because of the growing number of promising candidates but also because of their unique therapeutic potential (Senior, 2022). These biologics are highly diverse in terms of molecular size, resulting in different physical properties and different sensitivity to chemical and physical influences occurring during processing or storage. According to their variety in target and site of action, they require a wide range of dosage forms and thus manufacturing technologies. Additionally, due to the proteins' plethora, production quantities range from several kg/year up to 100 kg/year (Pennington et al., 2021). Respectively, the production of a final (personalized) dosage form for biologics is of interest on both large and small scales. Considering the storage stability under normal conditions, the transfer of proteins from their liquid state into dry, solid powders is advantageous.

To this end, we developed a new personalizable concept of directly dosing the protein solution (i.e., ink) via inkjet printing into the final vial, followed by a gentle drying step to end up in a stable dosage form. In our study, Human Serum Albumin (HSA, approx. 66kDa) stabilized in phosphate buffered saline was used as a model protein. Dispensing a single (personalizable) dose of protein into its final vial was achieved via an industrial inkjet printer. Subsequently, vacuum drying at ambient temperature and 50 mbar was applied.

All investigated DoE formulations were printable fluids, but within the *satellite droplets* regime ($Oh<0.10$, $40<Re<125$) (Derby, 2010). Thereby, we found PS80 to be crucial for reducing the jetted drop angle to allow for precise drop deposition at the vial bottom. Further, the residual moisture of the dried doses revealed acceptable water content (<10wt%) to maintain the protein flexibility and long-term stability. Therefore, the two optimized formulations were consistently printable and no increase in protein aggregation was detected when printing over prolonged time period. Further, 93% of a targeted dose were successfully dispensed into the final vial and reconstitution volumes down to 100µl resulted in no significant loss in reconstituted dose. To conclude, the developed concept is a promising approach to broaden the available processes for (personalizable) flexible production of solid dosage forms, allowing for fast adaption of production scales.

Literature:

Derby, B., 2010. Inkjet printing of functional and structural materials: Fluid property requirements, feature stability, and resolution. Annu. Rev. Mater. Res. 40, 395–414. https://doi.org/10.1146/annurev-matsci-070909-104502

Pennington, M.W., Zell, B., Bai, C.J., 2021. Commercial manufacturing of current good manufacturing practice peptides spanning the gamut from neoantigen to commercial large-scale products. Med. Drug Discov. 9, 100071. https://doi.org/10.1016/j.medidd.2020.100071

Senior, M., 2022. Fresh from the biotech pipeline: too much, too fast? Nat. Biotechnol. 40, 155–162. https://doi.org/10.1038/s41587-022-01208-2

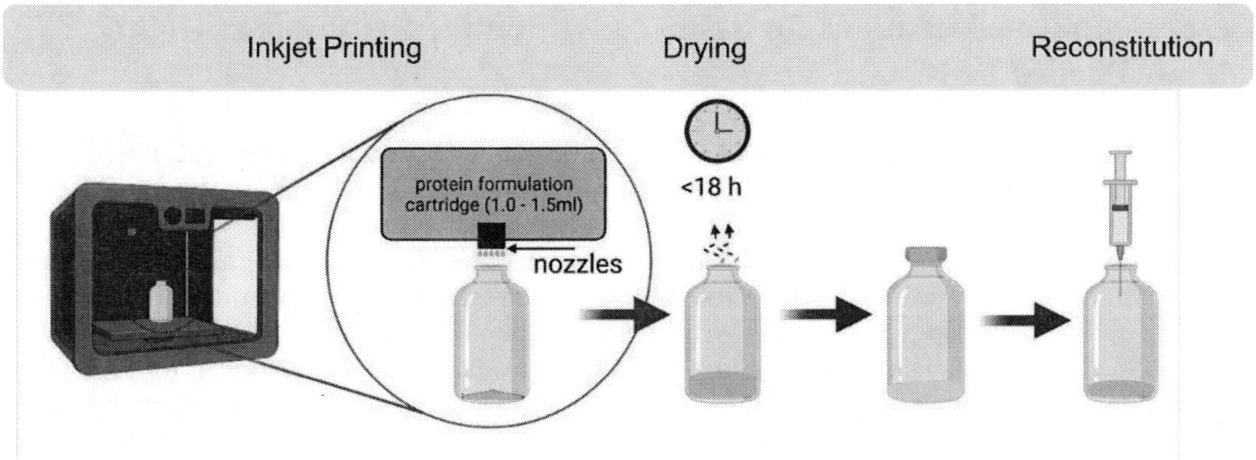

Inkjet Printing
Drying
Reconstitution
protein formulation cartridge (1.0 - 1.5ml)
nozzles
<18 h

605a Scaleup Considerations for Solid-Phase Peptide Synthesis in Highly Dynamic Packed Beds.

Harrison Rose, Mike Di Maso, Kyle E. Ruhl, Shane T. Grosser, François Lévesque, Kevin M. Maloney, U. Faruk Mansoor, Ritwika Ray, Danielle M. Schultz, Steven M. Silverman

Macrocyclic peptides are a growing and promising asset class within the development pipeline at Merck andamp; Co., Inc. (Rahway, NJ). While batch automated solid-phase peptide synthesis (SPPS) is the foundational technology for accessing milligrams of peptide, it does not align with Merck's Green and Sustainable Science guidelines for larger-scale production due to the need for high solvent volumes and stoichiometric excesses of non-canonical amino acids. These considerations suggest that scaleup efforts be directed toward developing a packed-bed reactor configuration which may offer improvements to volumetric productivity and process mass intensity. However, the resin comprising the solid phase tends to undergo dramatic and peptide-specific swelling and contraction over the course of each coupling, which necessitates careful consideration of scaling parameters, as well as development of novel control methods that are challenging to integrate with existing piloting and manufacturing process equipment.

To address these issues, Merck has identified key scaling parameters and developed dynamic axial compression variable bed reactors with closed-loop bed height control to provide line-of-site for more environmentally sustainable scaleup of SPPS to support kilo- and multi-kilo-scale deliveries of macrocyclic peptide APIs.

605b New Insights into the Difficulties and Importance of Dichloroacetic Acid Removal Post Detritylation in the Solid Phase Oligonucleotides Synthesis Cycle

Peng-Kai Kao, Todd Kajdan, Timothy M. Braden, Chen-Chun Chen, David Logsdon, Wei-Ming Sun, Eric D. Moher, Martin Johnson, Yufei Wei

Therapeutic oligonucleotides have received significant attention across the pharmaceutical industry. Solid phase oligonucleotide synthesis (SPOS) is the main approach of manufacturing oligonucleotides. SPOS is well known to have high cost and large amount of organic waste [1]. Organic solvent used in post-reaction washing is the main contributor to SPOS waste generation. In this study, we have investigated the impact of washing in SPOS. The importance of washing away dichloroacetic acid (DCA) after dimethoxytrityl (DMT) group deprotection and before coupling with the next activated phosphoroamidite is well known. However, the binding of DCA to the resin and the growing oligonucleotide, and the limiting factors of DCA removal from the resin bed prior to coupling have not been studied on a fundamental level. We have measured the adsorption-desorption isotherms of DCA on the solid phase at different progress of oligo synthesis. The longer the oligonucleotide strand on solid phase, the more DCA can be adsorbed on solid phase at certain concentrations of DCA in the liquid phase. Longer oligonucleotide strand also leads to more adsorption-desorption hysteresis, which significantly impacts the amount of solvent required for washing away DCA prior to coupling. Based on these findings, we have engineered washing strategies that yields high purity and high yield and significantly enhancing the sustainability of the oligonucleotide synthesis process.

References

[1] B. I. Andrews, F. D. Antia, S. B. Brueggemeier, L. J. Diorazio, S. G. Koenig, M. E. Kopach, H. Lee, M. Olbrich and A. L. Watson, 'Sustainability Challenges and Opportunities in Oligonucleotide Manufacturing,' J. Org. Chem, 2021, 86, 49-61.

605c Process Bottleneck of Packed Bed Oligonucleotide Reactors Arising from Critical Flux Due to Solid Phase Compressibility

Yufei Wei, Al Khilevich, Wei-Ming Sun, Martin Johnson, Timothy M. Braden, Chen-Chun Chen

Oligonucleotides have received growing attention as a new class of therapeutic. To date, over ten therapeutic oligonucleotides have hit the market and more than hundred are under clinical development. Some oligonucleotides under development have broad target patient population and are expected to have annual demand of over 1 metric ton. At present, solid phase oligonucleotide synthesis in packed bed reactors is the main manufacturing approach. The bed height is usually limited below 10 cm. Over-pressurization often occurs in the first 1 or 2 cycles when using polystyrene resin if running with a bed height that would eventually reach 10 cm by the end of the oligonucleotide build, which poses a major bottleneck for process throughput.

In this work, we have studied the pressure drop across a packed bed of polystyrene resin for oligonucleotide synthesis and quantified critical flux as a function of bed height, solvent, and length of oligo. We have found that the relationship between pressure drop and flux is nonlinear and the pressure drop increases exponentially as the flux reaches a certain critical value. We have also found that the packed bed is compressible under flow – bed height decreases as flux increases. Factors impacting the critical flux include the solvent, the bed height, the bed diameter and the length of oligonucleotide on resin. The more the solvent swells the resin, the lower the critical flux is in that solvent. The taller the bed, the lower the critical flux. The critical flux decreases as the bed diameter increases until it reaches around 2.5 cm. The longer the oligonucleotide on resin, the higher the critical flux (smaller specific cake resistance). We developed a mathematical model by considering bed compression under pressure drop and corresponding change in local porosity. This model can quantitatively describe our experimental data. We also developed a novel synthesis condition that allows detritylation reaction to be conducted in acetonitrile instead of toluene. There are already commercially available ACN-rich oxidation and sulfurization reagents. Performing detritylation in acetonitrile, together with those ACN-rich oxidation and sulfurization reagents, and with less concentrated capping reagents, over-pressurization problem can be made less severe, bed height can be higher, more solid phase can be packed in reactor, and therefore process throughput can be greatly increased.

605d Study of Factors Impacting Peptide Retention in Nanofiltration

Paridhi Agrawal, Kevin D. Seibert

Nanofiltration (NF) is a membrane separation unit operation which has applications across several industries such as water treatment, food and agriculture, fine chemicals, and pharmaceutics. It is used for concentration, purification by removal of low molecular weight species, and solvent exchange by applying high pressure across a nano-porous membrane orthogonal to the bulk flow. For the processing of larger molecules such as proteins, monoclonal antibodies, and peptides where intermediate or final isolation becomes challenging, membrane filtration can be a powerful tool. It also aligns well with other upstream and downstream continuous unit operations to make the overall synthesis telescoped and aids in improving process control and throughput. Membrane material and molecular weight cut off (MWCO) selection is done based on the physical and chemical properties of the solvents and solutes involved in the process.

The performance of an NF process is measured based on cycle time and yield. The parameters that impact these include feed flow and crossflow rate, transmembrane pressure (TMP), temperature, membrane pore size, and viscosity of solvent system. If chemistry modifications are not possible, the only controllable process parameters are flowrate and TMP. Being a high-pressure technology, one challenge in NF is the loss of solute molecules to the permeate even after selection of a tight membrane. This can be addressed by membrane conditioning using the process solution wherein a solute layer forms on the membrane which acts as a secondary sieve. Since this polarization layer adds resistance in the system, it causes reduction of permeate flux and increases cycle time. Operating conditions such as TMP and crossflow rate need to be adjusted to balance the flux reduction and yield loss.

In this work, we utilized ceramic NF membranes with an MWCO of 450 Da to study the effect of operating parameters on the retention of an ~2500 Da peptide molecule. TMP and crossflow rate were modified to span the flow regime spectrum from laminar to turbulent, and low to high pressure. Caustic regeneration was performed between each experiment to reset the membrane to the same baseline clean solvent flux. From experiments, we show that for this ratio of peptide size to membrane pore size, high crossflow velocity and low TMP aid in peptide retention.

605e Digital Process Replica of RNA Manufacturing for Enabling Process Optimization, Technology Transfer and Techno-Economic Assessment

Simon Daniel, Zoltán Kis, Cleo Kontoravdi, Nilay Shah

Methodology

Our approach starts with an extensive review of relevant, large-scale, cGMP RNA manufacturing processes, encompassing current Covid-19 vaccines production settings and potential innovations. This enables the identification of crucial process steps, critical process parameters, and relevant manufacturing options to be simulated (See **Figure 1**). In this study, particular emphasis is placed on the *In Vitro* Transcription (IvT) modes of operations and RNA 5'-capping options. Then, process models are built for each unit procedure to integrate IvT kinetic modelling with downstream and formulation simulations. This holistic modelling approach allows the prediction of the main key performance indicators (KPIs) and critical quality attributes (CQAs) under different scenarios. This includes RNA yield, recovery, process time, production cost, RNA degradation, and the formation and removal of product-related impurities throughout the manufacturing process. This model-based assessment is however limited by the lack of transparency, as well as by inherent variabilities and unknowns surrounding process design, equipment, and location. A sensitivity analysis is thus performed to embrace these uncertainties. A variance-based global sensitivity analysis has been selected to perform 10,000 quasi-random sampling based on low-discrepancy Sobol sequences. These simulations are used to explore the input-output uncertainty propagation and compare the different manufacturing options in a risk-based manner. Random Sampling-High dimensional Model Representation (RS-HDMR) metamodeling is also implemented to obtain an accurate estimation of the sensitivity indices, indicating the degree to which these variables influence process performance and product quality (see **Figure 2**).

Results

First, process simulations confirm the high production capacity of the RNA technology. The equivalent of hundreds of millions of Covid-19 vaccines doses could be produced each year by a single production line. However, the sensitivity analysis reveals significant discrepancies. The LNP formulation step is identified as the production bottleneck in most cases. Thus, the choice of mixing method, device parallelization and numbering is critical to ensure optimal annual productivity, and is highly dependent on the IvT mode of operation and expected yield. Secondly, in sharp contrast with other biopharmaceutical productions, the consumptions of IvT raw materials and capping agent are the main cost drivers. For this reason, the fed-batch approach is economically attractive, especially in combination with co-transcriptional capping. The trade-off, however, is an additional production of RNA-related impurities, further increasing downstream burden to obtain high quality products. Similarly, the application of high-temperature conditions can make the enzymatic capping approach cost competitive. Ultimately, the transition from batch to continuous operations is highly attractive in terms of process transferability and product quality. The optimization of process occupancy would enable the development of a smaller and cheaper facilities. IvT reactor scale can be at least divided by 10, further reducing the footprints of all other downstream units. Process streamlining can finally lead to raw materials recycling, making the vaccine supply chain more robust and the process more cost-efficient

Implications and future perspective

This work represents the first end-to-end, integrated modelling of RNA manufacturing processes encompassing different manufacturing options. This process digitization is particularly useful in such nascent platform technology. It provides a reliable tool for assessing the requirements for setting up a new facility and can help would-be manufacturers establish a suited manufacturing strategy. This analysis also points out where process intensification and development are most needed in order to optimize manufacturing capabilities, process transferability, production costs and product quality. We further showcase that continuous manufacturing is technically feasible and economically competitive. This switch to continuous operations could pave the way towards the deployment of a transferable, distributed, and affordable RNA platform technology. In another perspective, the identified design space and modelling capabilities could be at the basis of future model-based design of experiments and process design. Finally, these models could be expanded and complexified in the near future. The integration of process analytical technology (PAT) tools would notably increase our prediction capability and support the use of model-based process control.

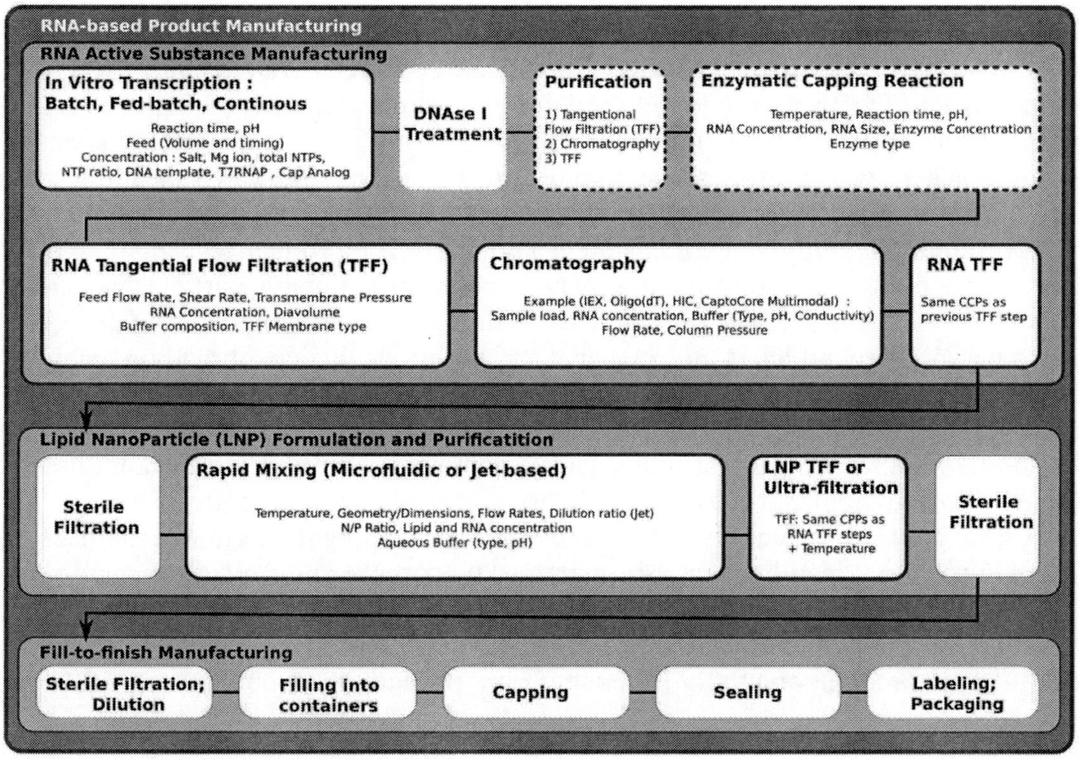

Figure 1 : Process steps and associated process parameters included in the digital process models of RNA manufacturing

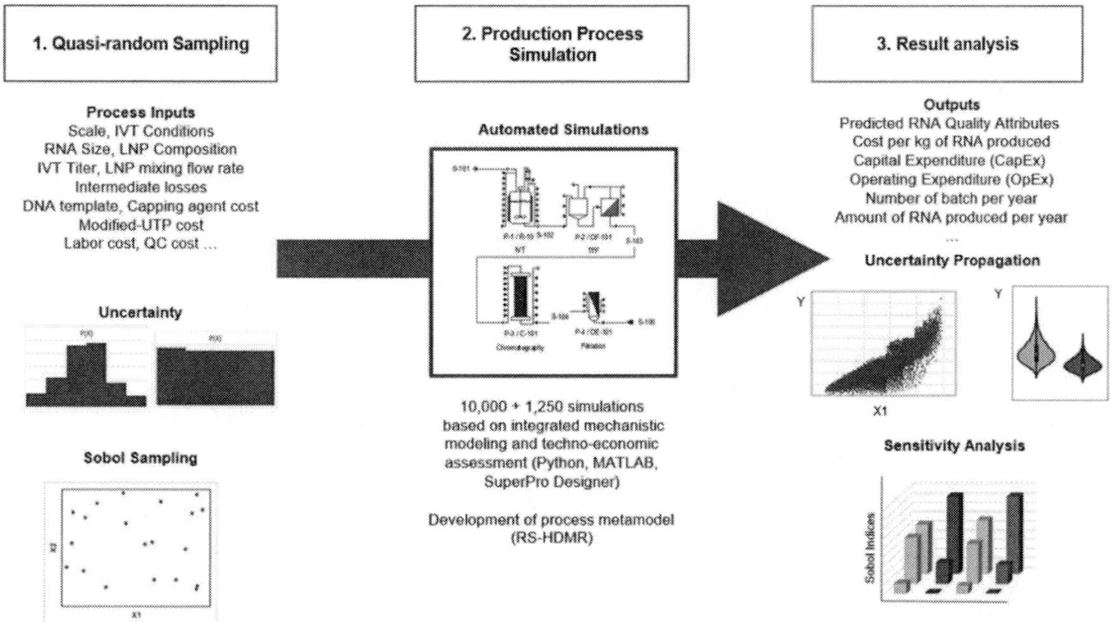

Figure 2 : Computational workflow for the model-based uncertainty and global sensitivity analyses

605f A 3D-Printed Microfluidic Device for High-Throughput Production of mRNA-Lipid Nanoparticles

Wan-Zhen Lin, Kristian Bostic, Noah Malmstadt

Lipid nanoparticles (LNPs) play a vital role in drug delivery as seen in the recent life-saving messenger RNA (mRNA) vaccines against coronavirus disease 2019 (COVID-19), mRNA-1273 and BNT162b. The entrapment of mRNA is accomplished during particle formation, when the cationic lipids interact with anionic mRNA. Production of LNPs via microfluidic hydrodynamic focusing has been thoroughly investigated, where a focused stream of lipids is generated by two sheathing streams of aqueous buffer containing the mRNA. LNPs are formed as the lipids are rapidly mixed with the aqueous buffer. Microfluidic hydrodynamic focusing is a popular LNP production technique due to its unique advantages: 1) continuous-flow production, 2) controllable and tunable particle size, and 3) single-step fabrication. Despite its success, current microfluidic hydrodynamic focusing devices are fabricated by elastomer molding, which is an expensive and laborious device fabrication technique. The resulting devices often have poor user interfaces, are prone to clog due to the small channels, and are made of delicate materials that can only be used at low flow rates (μL/min range) which greatly limits the production throughput. To address these issues, we redesigned the hydrodynamic focusing devices by replacing the original planar channels with three-dimensional flow paths. By stereolithography printing, we fabricated a microfluidic device that is made of durable methacrylate-based materials and implements readily adaptable interfaces by commercial fluidic connectors. We demonstrated encapsulation of 4knt mRNA and characterized the resulting LNPs. With larger microchannels than conventional microfluidics, our 3D-printed microfluidic device can be used at higher flow rates (mL/min range), potentially increasing LNP production throughput by > 30-fold.

641a Comparison of One Dimensional and Two-Dimensional Population Balance Model for Optimization of a Crystallization Process

Bhavik Mehta, PhD, Niall Mitchell, Cameron Brown, Sebastion Davidson, Alastair J. Florence

Introduction:

Crystallization is a key unit operation in the isolation of many active pharmaceutical ingredients. The crystallization process commonly has a significant impact on the quality of drugs and the efficiency of downstream processes such as filtration, milling, centrifugation, drying, granulation, and tableting (Eren et al., 2019). One of the key challenges in industry is the integration of Quality-by-Design methods when designing unit operations and processes. This would help to provide potential ways to improve quality while reducing variability and is being actively encouraged by pharmaceutical regulatory agencies (Allison et al., 2015). The two-dimensional population balance model (2D PBM) allows for the modelling of two different growth kinetics, along perpendicular major andamp; minor axes of a particle. This approach has several advantages for the modelling crystallization processes that generate non-spherical particles, such as needles and plates, commonly encountered for active pharmaceutical ingredients (API) (Rosenbaum et al., 2022). The 2D PBM approach is also able to account for morphological evolution during a crystallization process; something that cannot be successfully accomplished with a 1D model.

Results:

This work outlines a stepwise approach to configuring, validating and optimizing a 2D PBM model. As a first step, the experimental data gathered is used to configure and validate a 1D PBM model. The initial validated 1D model showed that while the desupersaturation behaviour is well predicted and can be optimized using a 1D approach, the particle size did not match the PSD quantiles observed experimentally as shown in Figure 1 (attached).

From these findings, it is evident that the particle shape changes during the crystallization process, which results in a change in particle shape, something that the 1D approach is not able to account for. Therefore, to investigate this further, a 2D PBM model is configured and validated using the estimated model parameters from the 1D approach as initial guesses. The results indicate that the 2D approach is better equipped to predict the particle shape evolution during the process, as shown in Figure 2 (attached)

These results were used to explore the design space and optimize the crystallization process with a view of identifying process parameters which may results in an improved product particle, namely lower aspect ratio (major axis/minor axis).

References

Allison, G., Cain, Y. T., Cooney, C., Garcia, T., Bizjak, T. G., Holte, O., . . . Zezza, D. (2015). Regulatory and quality considerations for continuous manufacturing. May 20-21, 2014 Continuous Manufacturing Symposium. *J Pharm Sci, 104*(3), 803-812.

Eren, A., Szilagyi, B., Quon, J., Furuta, M., Papageorgiou, C., Nagy, Z., . . . Ozkan, L. (2019). Development of a Model-Based Quality-by-Control Framework for Crystallization Design. *29th European Symposium on Computer Aided Process Engineering, Pt a, 46*, 319-324. doi:10.1016/B978-0-12-818634-3.50054-0

Rosenbaum, T., Mbachu, V., Mitchell, N. A., Gamble, J. F., Cho, P., andamp; Engstrom, J. D. (2022). Comparison of One-Dimensional and Two-Dimensional Population Balance Models for Optimization of a Crystallization Process for a Needle-Shaped Active Pharmaceutical Ingredient. *Organic Process Research andamp; Development.* doi:10.1021/acs.oprd.1c00344

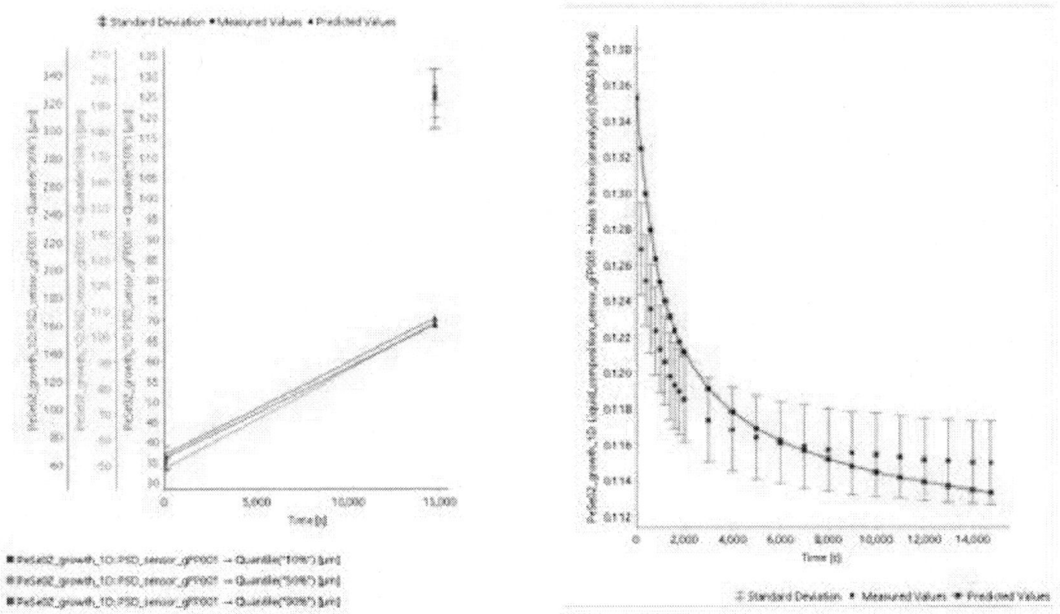

Figure 1: 1D PBM Crystallization model predictions

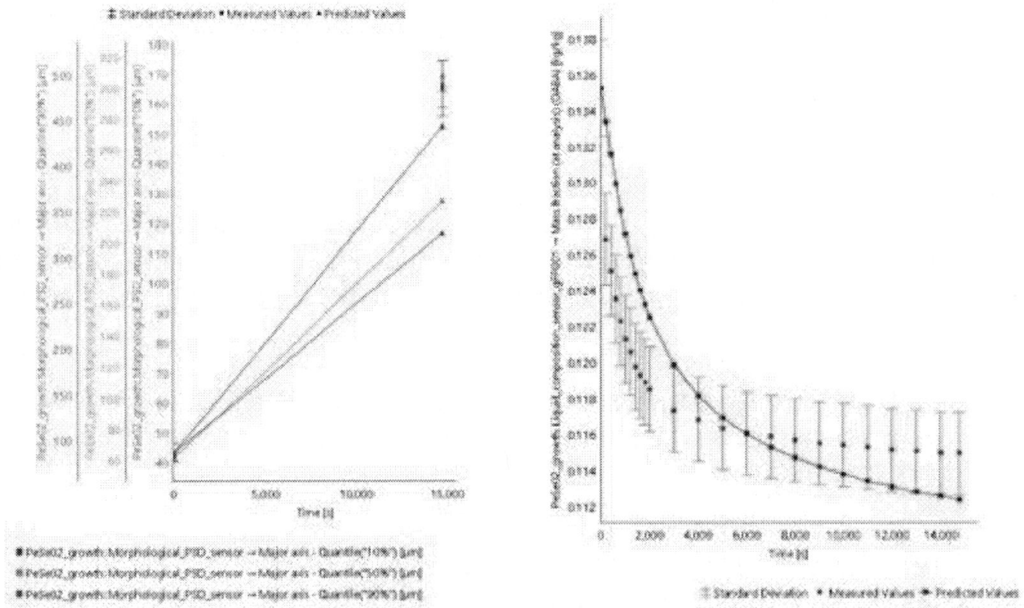

Figure 2: 2D PBM Crystallization model predictions

641b In-Depth Analysis of the Impacts of Material Properties on Size Distributions in Continuous Twin-Screw Wet Granulation to Construct a Generic 1D Population Balance Model

Kensaku Matsunami, Ana Alejandra Barrera Jimenez, Michiel Peeters, Daan Van Hauwermeiren, Thomas De Beer, Ingmar Nopens

Continuous twin-screw wet granulation (TSWG) is a key unit operation for the continuous manufacturing of several solid oral dosages. To understand the mechanism of TSWG deeply, numerous efforts have been made based on experimental approaches. For example, the liquid-to-solid (L/S) ratio has been identified as a critical process parameter for granule quality attributes[1]. The knowledge obtained by experimental studies can help the process development for new formulations qualitatively.

In addition to experimental approaches, mechanistic modeling has been widely performed to predict granule quality attributes as well as to explain the phenomena mathematically. Population balance models (PBMs) are popular methods to compute granule particle size distributions (PSDs) in TSWG. So far, multiple research groups have proposed PBMs with different dimensions and kernels[2–6]. One of the biggest challenges of PBMs in TSWG is the applicability for different formulations containing new active pharmaceutical ingredients (APIs). While material properties of APIs are variable, e.g., from hydrophilic to hydrophobic materials, most of the PBM studies have focused on specific materials and are hence not generic.

This study presents an in-depth analysis of the impact of material properties in TSWG to construct a generic 1D PBM. A generic PBM can predict granule PSDs of different formulations and process conditions only by the information of material properties as well as process conditions. As a basis of it, a hybrid model of PBM and partial least squares (PLS) was developed to link material properties with granule PSDs. A compartmental one-dimensional PBM developed by the authors' research groups[7, 8] was calibrated for ten formulations, including seven different APIs with the same fillers and binders. Afterward, the impact of 34 material properties of the seven different APIs as well as L/S ratio on PBM parameters were trained by developing PLS models for each compartment. The hybrid model, linking the PBM and PLS models, was validated by applying it for four new formulations. A sensitivity analysis was performed to determine critical material properties as well as PBM parameters.

The validation results show that the proposed generic model could predict granule PSDs for new formulations in terms of the maximum mean discrepancies, which represents a statistical distance between two distributions. The maximum mean discrepancies between the simulated and the measured results for 18 experiments were between 0.031 and 0.143. General trends of the impact of material properties on PBM parameters were obtained through the variable importance for the projection of the PLS models. Size and moisture-related material properties affected the first wetting zone, whereas density-related material properties had a high impact on the subsequent kneading zones. From the results of sensitivity analysis, the aggregation efficiency at the wetting zone was found to be the most critical PBM parameter. In addition, analytical results showed that moisture-related properties, e.g., solubility, were key material properties to predict granule PSDs.

In conclusion, the developed model has a potential to reduce experimental work in the process development of new oral solid dosages. The hybrid generic model could predict granule PSDs for new

APIs when applying different L/S ratios by characterizing material properties, without any additional experiments. Furthermore, the importance of material properties was prioritized by sensitivity analysis, which decreased the number of the necessary material characterization methods for new APIs.

Acknowledgment: The authors would like to acknowledge, in no particular order, Janssen Pharmaceutica, Pfizer Inc., and UCB for their financial support and fruitful collaboration in this project.

References:

1. C. Portier, C. Vervaet, V. Vanhoorne, Continuous twin screw granulation: A review of recent progress and opportunities in formulation and equipment design, Pharmaceutics 13 (2021) 668.

2. K. F. Lee, S. Mosbach, M. Kraft, W. Wagner, A multi-compartment population balance model for high shear granulation, Computers \andamp; Chemical Engineering 75 (2015) 1–13.

3. H. Y. Ismail, M. Singh, A. B. Albadarin, G. M. Walker, Complete two dimensional population balance modelling of wet granulation in twin screw, International Journal of Pharmaceutics 591 (2020) 120018.

4. C. Sampat, Y. Baranwal, R. Ramachandran, Accelerating multidimensional population balance model simulations via a highly scalable framework using GPUs., Computers \andamp; Chemical Engineering (2020) 106935.

5. L. G. Wang, S. U. Pradhan, C. Wassgren, D. Barrasso, D. Slade, J. D. Litster, A breakage kernel for use in population balance modelling of twin screw granulation, Powder Technology 363 (2020) 525–540.

6. L. G. Wang, J. P. Morrissey, D. Barrasso, D. Slade, S. Clifford, G. Reynolds, J. Y. Ooi, J. D. Litster, Model driven design for twin screw granulation using mechanistic-based population balance model, International Journal of Pharmaceutics 607 (2021) 120939

7. D. Van Hauwermeiren, M. Verstraeten, P. Doshi, M. T. am Ende, N. Turnbull, K. Lee, T. De Beer, I. Nopens, On the modelling of granule size distributions in twin-screw wet granulation: Calibration of a novel compartmental population balance model, Powder Technology 341 (2019) 116–125.

8. A. A. Barrera Jiménez, D. Van Hauwermeiren, M. Peeters, T. De Beer, I. Nopens, Improvement of a 1d population balance model for twin-screw wet granulation by using identifiability analysis, Pharmaceutics 13 (2021) 692.

641c Use of gPROMS for Film Coating Processes - Streamlined QbD Process Development

James Miesle, Salvador Salvador Garcia

The film coating process is ubiquitous across the pharmaceutical industry and is an essential step in the manufacture of tablets. Film coating is desirable for numerous reasons, including for functional purposes (i.e. modified or delayed release), product identity/branding, aiding in patient ingestion, taste masking, containment of the active substance, and facilitating downstream packaging operations. Numerous process parameters are inter-related in the film coating process, thereby impacting the final coating appearance and quality. These parameters generally impact one or more of the following areas: the thermodynamic environment, the mechanical movement of the tablets, and the spray application. This presentation outlines an approach to identify a common film coating process space, that can be used as a target operational space for new products. Based on existing commercial processes, through the use of global sensitivity analysis in gFormulatedProducts and a model for the film coating process. Two common process envelopes were generated for the most common film coating polymer systems (HPMC and PVA). Application of this approach has the potential to save both time and material, avoiding the potentially lengthy and typically iterative process development cycle.

641d Towards Optimized Pharmaceutical Freeze-Drying Processes Via Controlled Nucleation: Mechanistic Modeling of Vacuum-Induced Surface Freezing

Leif-Thore Deck, Andraž Košir, Marco Mazzotti

Keywords: Freezing, Drying, Freeze-Drying, Lyophilization, Stochastic Nucleation, Mechanistic Modeling, Pharmaceutical Manufacturing, Quality by Design, Process Design andamp; Development,

Motivation and Introduction

Freeze-drying is the method of choice to stabilize sensitive biopharmaceutical drug products, since it removes water at low temperatures. At commercial scale, tens of thousands of vials containing the pharmaceutical formulation are densely packed on temperature-controlled shelves and freeze-dried together. [1-2] Regulatory requirements impose that the products in all vials have to meet quality specifications, necessitating a detailed process understanding and tight control.

At the same time, the properties of dried products vary among vials on a shelf, rendering process design and control challenging. [1-3] This batch heterogeneity is expressed in a plethora of product attributes, whereby the mean pore size and the residual moisture in the dried products are of special interest. A variability in these properties directly poses a risk to the quality of the final product and thus necessitates the choice of more conservative process parameters, connected to longer cycle times. [1-2]

Literature indicates that the stochasticity of ice nucleation is a major cause for batch heterogeneity. The times and temperatures, at which nucleation occurs, inherently vary among vials; since the nucleation temperature is understood as key predictor of the mean pore size of the product, this variability affects the final properties of the dried product, including residual moisture. [1-3]

To overcome this stochasticity, a number of technical approaches have been developed over the past decades, commonly referred to as controlled nucleation techniques. [1-2] While many such techniques have been proposed, only three of them have emerged so far in laboratory scale freeze-drying devices, currently moving towards implementation at commercial scale. [3]

Among these techniques, vacuum-induced surface freezing (VISF) is the only one that does not require major technical modifications or retrofitting of the existing freeze-drying equipment and is thus of special interest for commercial applications. The technique is based on applying vacuum during the freezing stage, in order to promote nucleation via evaporative cooling. [3] While successful implementations of VISF were reported in literature, no systematic and mechanistic framework exists to date that could predict the process performance and thus guide the choice of operating conditions. And indeed, there are many parameters that were shown to affect the process outcome: They include the duration during which vacuum is applied and the corresponding pressure; the temperature of the shelf at the time of the vacuum; and finally, the design of holding steps at constant temperature both before and after the vacuum phase.

Given the plethora of process parameters, the choice of parameter values is primarily empirical, rendering process design and optimization challenging. This work aims to support lyophilization professionals in designing and understanding VISF by providing a mechanistic model of the process.

Methods

We extended a mechanistic freezing model recently developed in our group, [4-5] in order to study vacuum-induced surface freezing in detail. The model explicitly considers the stochastic nature of nucleation, and predicts the thermal gradients that form within the vial upon application of vacuum due to evaporation. Importantly, the effect of these gradients on the nucleation rate is taken into account. To promote the use and further development of the model in the wider community, we provide open source access to it in the form of a python package (available at https://pypi.org/project/ethz-snow/).

We applied the model to carry out sensitivity studies to deepen the understanding how various process parameters affect the process performance. Additionally, we compared the model predictions with experimental data from campaigns in laboratory freeze-driers.

Results

Model instantiations revealed that the extent of evaporative cooling induced by the application of vacuum increases for longer vacuum duration and lower vacuum pressure. The vacuum pressure was found to govern the final minimum temperature that could be achieved. In order to promote nucleation efficiently, low temperatures have to be reached fast, i.e. before a substantive amount of water has evaporated. For this, the application of lower pressures was found to be beneficial.

In addition to process conditions, the nucleation behavior of a formulation is governed by its nucleation kinetics. We therefore studied the effect of the nucleation kinetics on the performance of VISF; the findings indicate that the process becomes less efficient for slower nucleation kinetics, where lower temperatures have to be reached for nucleation to occur.

Finally, we studied the effect of holding steps on the VISF outcome. In line with the current practice in experimental case studies, we find that holding steps both before and after the application of vacuum are beneficial. A pre-nucleation holding step ensures thermal equilibration; a post-nucleation holding step ensures that solidification takes place in a slow and controlled manner. To this end, the model identifies suitable temperatures and durations for both holds.

Discussion and Implications

The finding that the performance of VISF depends on the nucleation kinetics of the specific formulation, is of great relevance. It implies that the choice of process parameters requires knowledge on the nucleation behavior of the formulation; different formulations may require different process conditions to achieve optimal process performance.

For a given formulation, nucleation occurs in average at lower temperatures the higher the concentration is. Thus, VISF eventually reaches performance limitations for highly concentrated systems, where the extent of evaporative cooling is not sufficient anymore to induce nucleation. Furthermore, the nucleation kinetics is governed by the number of available nucleation sites, which depend on the production environment: Under GMP conditions, reported nucleation temperatures typically are 5-10 K lower than those obtained for the same formulation under standard laboratory conditions. This implies a potential scale-up issue, where VISF parameters successfully tested in experiments at lab-scale may be unsuitable for commercial-scale processes. The model may be used to quantify this difference and suggest how parameter values have to be adjusted depending on the environment.

As these findings indicate, the herein proposed model provides a valuable framework to study VISF in a mechanistic fashion. It may be used to understand potential limitations of the process, and to identify suitable process conditions tailored towards the specific formulation of interest.

Acknowledgement

The authors thank the Janssen Pharmaceutical Companies of Johnson andamp; Johnson for the support in the course of the project and the financial funding. We especially thank Juan Carlos Araque and Ryan Wall.

References

[1] JC Kasper and W Friess: The freezing step in lyophilization: Physicochemical fundamentals, freezing methods and consequences on process performance and quality attributes of biopharmaceuticals, *Eur. J. Pharm. Biopharm.* (2011), 78, 248-263.

[2] LT Deck, DR Ochsenbein and M Mazzotti: Stochastic Shelf-Scale Modeling Framework for the Freezing Stage in Freeze-Drying Processes, *Int. J. Pharm.* (2022), 613, 121276.

[3] R Pisano, A Arsiccio, K Nakagawa and AA Barresi: Tuning, measurement and prediction of the impact of freezing on product morphology: A step toward improved design of freeze-drying cycles, *Drying Technology* (2019), 37:5, 579-599.

[4] LT Deck, DR Ochsenbein and M Mazzotti: SNOW – Stochastic Nucleation of Water, (2021), GitHub Repository, https://github.com/SPLIfA/snow/

[5] GM Maggioni and M Mazzotti: Modelling the stochastic behaviour of primary nucleation. *Faraday Discussions* 2015, 179, 359-382.

641e Modelling the Stress Distribution in Tablet Coatings Exposed to Rapid Environmental Changes

Venugopala S. Punati, Mahesh S. Tirumkudulu, Ashwinkumar Jain, Daniel O. Blackwood, Alfred Berchielli, Pankaj Doshi

Non-functional film coatings are applied to core tablets to improve product appearance, impart taste-masking properties, improve handling and stability of the dosage form, and reduce exposure to active drug substance for caregivers. The coatings protect tablets from negative effects of environment such as humidity, oxidation, and light. The mechanical stability and integrity of tablet coatings are therefore important to maintain drug product quality attributes such as appearance and stability. Therefore, environmental conditions under which these coatings may crack are important to understand so as to prevent their occurrence. In this work, we present a novel computational framework to assess the mechanical integrity of tablet coatings exposed to rapid variations in environmental conditions. We perform detailed stress and strain analysis of tablet coatings on tablet surfaces with debossed regions and identify conditions for cracking for both elastic and viscoelastic coatings. Rapid changes in environmental temperature and humidity can cause differential expansion/contraction of coating and tablet core resulting in stresses that are many times more than those experienced during the drying process in a coater. Debossed regions on the tablet surface with sharp surface curvatures act as stress concentrators that may lead to coating cracks. Small changes in the design of the debossed regions lead to modest reductions in the peak stress. Stress calculations suggest that coating crack is much more likely when the coating delaminates from the surface of the tablet core.

The attached figure presents the stress distribution for a bi-convex tablet with the letter `0' debossed on the top surface of the tablet. (a) Maximum in-plane principal stress in the coating of the tablet with 0.24mm debossed depth and 0.2mm corner radius in the debossed letter. (b) Coating stresses after reducing the debossing depth to 0.14mm, while keeping the corner radius constant. (c) Coating stresses after increasing the corner radius to 0.4mm keeping the debossing depth at 0.24mm. (d) Coating stresses in the tablets with more rounded letter (or oval '0') with a debossing depth of 0.24mm.

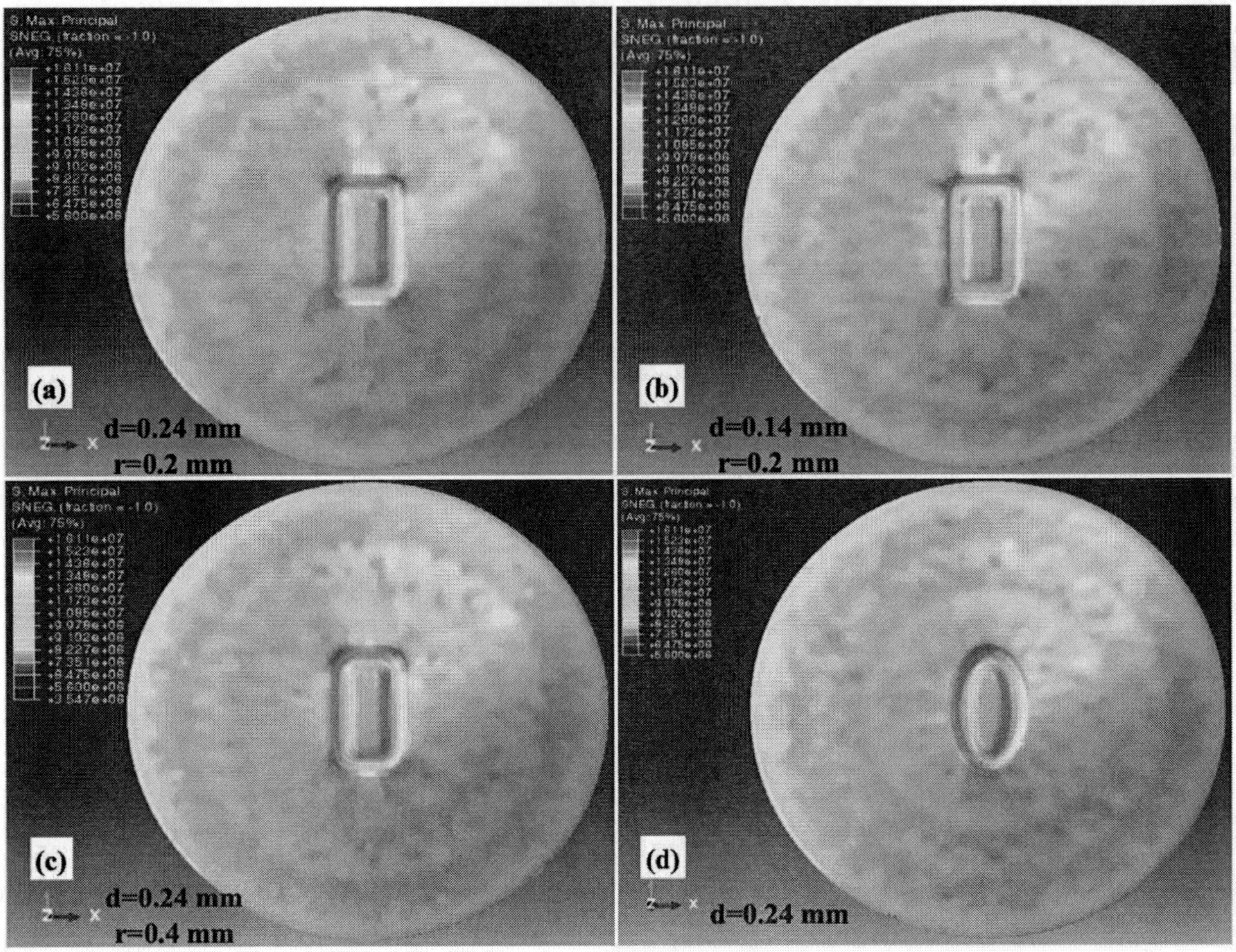
(a)
d=0.24 mm
r=0.2 mm
(b)
d=0.14 mm
r=0.2 mm
(c)
d=0.24 mm
r=0.4 mm
(d)
d=0.24 mm

641f Modeling the Steady State of a Spray Drying Unit with Drying Gas Recirculation Accounting for Complex Thermodynamics Using Sequential Simulation

Salvador Salvador Garcia

Solid dispersions are a way to aid the solubilization of molecules that are poorly soluble in water, and hence make them a candidate to be dosed orally. These intermediate materials are manufactured by spray drying a mixture of a chosen solvent(s), a polymer, and the drug [1]. Several modeling approaches have been proposed to address the drying dynamics in the spray drying unit. Ranging from the empirical model approaches [1] to the use of dimensional and dimensionless numbers [2], to bulk modeling approaches [3] to detailed models that incorporate the droplet level mechanisms [4] and complex measurement systems [5].

The reality of the industrial practice of developing a spray drying process is far from the well instrumented experimental settings in academia. The only real-time measurements available are those of the bulk (for both process and materials). Benchtop tests provide data of the state of materials at equilibrium (e.g. adsorption data generated by a dynamic vapor sorption instrument) however these will not account for the kinetics. As of the time of this work, acoustic droplet levitation technology remains mostly an academic asset and is not a common testing instrument found in industry. As such, the kinetics of drying (although important for particle morphology) remain an unobservable phenomenon given the bulk measurements available. Additionally, from a regulatory standpoint, the definition of process parameters for spray drying (operating conditions the process would be registered at) refer to static conditions representative of the steady state, assuming the process is in a state of control.

In the absence of particulate level information, practice is to control the environmental conditions of the process, with the assumption that keeping these in control will result in product quality control. This is common in granulation and coating processes where thermodynamic conditions as used as a surrogate to controlling the unobservable phenomena happening at the particulate or droplet level. This thermodynamic environment in the drying chamber of a spray drying unit is one of the main control mechanisms for the drying phenomena (aside from the atomization) and requires careful consideration of the way the physical properties of the gas phase are calculated.

This work presents the development and use of a steady state model for spray drying using Aspen Plus (Aspen Technology Inc.), accounting for complex thermodynamics and accommodating the different control schemes in the units. The details of the development of the flowsheet will be described, including the adjustments of the numerical method used by the simulator, necessary to solve the flowsheet.

[1] Friesen, D.T., Shanker, R., Crew, M., Smithey, D.T., Curatolo, W.J. and Nightingale, J.A.S., 2008. Hydroxypropyl methylcellulose acetate succinate-based spray-dried dispersions: an overview. *Molecular pharmaceutics*, *5*(6), pp.1003-1019.

[2] Garcia-Munoz, S. and Settell, D., 2009. Application of multivariate latent variable modeling to pilot-scale spray drying monitoring and fault detection: monitoring with fundamental knowledge. *Computers andamp; Chemical Engineering*, *33*(12), pp.2106-2110.

[3] Dobry, D.E., Settell, D.M., Baumann, J.M., Ray, R.J., Graham, L.J. and Beyerinck, R.A., 2009. A

model-based methodology for spray-drying process development. *Journal of pharmaceutical innovation*, *4*(3), pp.133-142.

[4] Oakley, David E. 'Spray dryer modeling in theory and practice.' *Drying Technology* 22, no. 6 (2004): 1371-1402.

[5] Schiffter, H. and Lee, G., 2007. Single-droplet evaporation kinetics and particle formation in an acoustic levitator. Part 1: Evaporation of water microdroplets assessed using boundary-layer and acoustic levitation theories. *Journal of pharmaceutical sciences*, *96*(9), pp.2274-2283.

[6] Radnik, J., Bentrup, U., Leiterer, J., Brückner, A. and Emmerling, F., 2011. Levitated droplets as model system for spray drying of complex oxides: A simultaneous in situ X-ray diffraction/Raman study. *Chemistry of Materials*, *23*(24), pp.5425-5431.

641g A Novel Adaptive Residence Time Distribution (RTD) Modelling Toolbox

Ravendra Singh, Fernando Muzzio

Residence time distribution (RTD) is a probability distribution function that describes how long a fluid or powder element spends inside a given operation. Currently, continuous pharmaceutical manufacturing is facing several challenges such as prediction of mixing in each unit operation as well as in an integrated line, assurance of content uniformity (CU), and tracing of materials for lots and batches identification ideally needed for product regulation and urgent re-calls [1-4]. These issues are directly related with patient safety and therefore are critical to address. The RTD model has been found in the center of all solutions to address these critical issues. However, because of different level of complexities, the accurate development and effective applications of the RTD model is still a challenging task. Therefore, a systematic toolbox is needed through which, the RTD model can be developed and deployed for varies applications.

In this work, an RTD toolbox has been developed for continuous pharmaceutical manufacturing process. The components of RTD toolbox includes, unit operation building blocks, unit operation RTD model library, calculation systems to perform varies calculations needed to develop integrated RTD based flowsheet model, miscellaneous models, and a library consisting of the integrated RTD based flowsheet model of varies manufacturing lines. The mathematical models have been developed to predict the RTD model parameters from operating conditions. The RTD toolbox is adaptive in nature and thus can be used to develop the RTD model of any continuous pharmaceutical tablet manufacturing process.

The applications of the RTD toolbox to develop the integrated flowsheet model of continuous pharmaceutical tablet manufacturing process has been demonstrated. The integrated flowsheet model is capable of predicting the mixing in each unit operation as well as its affects in final product. The model library and knowledge base are two important supporting tools for the developed toolbox. The model library consists of the RTD model of basic elements such as plug flow process, continuous stirred tank (CST), and dead zone (DZ). Subsequently, via combining these basic elements, the pharmaceutical unit operation (e.g. blender, feed frame) RTD model library has been developed. The knowledge base which has been developed through extensive experimentations consist of the RTD model parameters for a broad range of equipment's. This knowledgebase is used to develop the additional mathematic models to predict the RTD parameters. The integrated flowsheet model of any continuous pharmaceutical manufacturing processes can be easily generated using this toolbox via drag and drop approach. The toolbox is flexible in nature, can be adapted for any continuous pharmaceutical manufacturing plant and thus should have a broad range of real time applications through distributed control system (DCS) as well as for offline research and development purposes.

The varies applications of the RTD based integrated flowsheet model has been demonstrated. Some of these applications are as follows: prediction of API composition throughout the continuous pharmaceutical manufacturing process. Analysis of effects of disturbances and measurement noise. Material traceability. The applications of the RTD based digital twin for real time diversion of tablets have been also demonstrated to assure the content uniformity of the tablets.

The objective of this presentation is to folds. First to highlight the development of the 'adaptive RTD toolbox' and then to demonstrate its applications for development of integrated flowsheet model and thereby prediction of mixing in continuous pharmaceutical manufacturing, and assurance of content uniformity via real time tablet diversion.

References

1. Singh, R. (2021). Predicting tablet potency in continuous manufacturing. A modular toolbox enables residence time distribution-based control for continuous pharmaceutical manufacturing. Pharmaceutical Technology. 44-48.
2. Singh, R. (2019). Systematic framework for implementation of RTD based control system into continuous pharmaceutical manufacturing pilot-plant. Pharma. Issue 34, 43-46.
3. Bhaskar, A., Singh, R. (2018). Residence time distribution (RTD) based control system for continuous pharmaceutical manufacturing process. Journal of Pharmaceutical Innovation. DOI: 10.1007/s12247-018-9356-7.
4. Billups, M., Singh, R. (2018). Systematic framework for implementation of material traceability into continuous pharmaceutical tablet manufacturing process. Journal of Pharmaceutical Innovation. https://doi.org/10.1007/s12247-018-9362-9.

673a Exploring the Effects of Equipment Operational Defects on the Dissolution Performance in USP Apparatus 1

Eric Murphy, Julie Spence, Daniel Reinders, Eric Nielsen, Susan George, Yi Gao, Ann M. Czyzewski, Nandkishor Nere

Dissolution methods remain an important quality control tool for ensuring the safety and efficacy of tablets and capsules in the pharmaceutical industry. The quality of the drug product is ensured through the comparison of dissolution profiles over the course of time. Despite the use of standardized equipment and protocols, systematic differences in dissolution profiles sometimes arise between different apparatus of the same USP type. These differences may also arise inconsistently depending on operating variables such as shaft rpm. Delays from uncovering the culprit of these differences can produce serious costs when bringing new formulations to market. In this work, we examine how operational variation such as vibrations and wobble in a shaft manifest in differences in dissolution profiles by influencing the fluid mechanics. Specifically, this work investigates the USP Apparatus 1. We utilize computational fluid dynamics (CFD) modeling to quantify the effects of these defects on local turbulence characteristics. Standard dissolution equations could then be used to explore the sensitivity of dissolution profiles, thereby establishing a link between the observed dissolution performance and the underlying fluid mechanics. The results of this work are expected to provide a rational basis for the discrepancies observed by dissolution practitioners.

Disclosure: Authors are employees of AbbVie and may own AbbVie stock. AbbVie sponsored and funded the study; contributed to the design; participated in the collection, analysis, and interpretation of data, and in writing, reviewing, and approval of the final publication

673b First Principles and Cheminformatics Based in-Silico Tools for Physical Property Prediction Towards Informed Process Development

Sharad Maheshwari, Pelin Su Bulutoglu, Eric Murphy, Kushal Sinha, Akshay Korde, Moussa Boukerche, James Marek, Ryan Ellis, Manish Kelkar, Daniel Pohlman, Richard Hong, Nathan Abraham, Rajni Miglani Bhardwaj, Jeremy Henle, Nandkishor Nere

First-principles and cheminformatics based in-silico tools have recently shown great promise to accelerate drug development pipelines[1-5]. Using computational approaches for physical property predictions renders great utility both in early stages of molecular discovery and later stages of product formulation and process development. Solubility is one of the critical properties for design of various unit processes and operations including chemical reactions, extractions and more importantly crystallization processes for synthesis and purification. To characterize the performance of the various unit operations, it is often necessary to measure solubility of a compound in multiple solvents and mixture of solvents as a function of temperature. Thus, many solubility evaluations are required to enable multiple synthetic and physical transformations required to make active pharmaceutical ingredient (API). In this work we present a computational solubility prediction tool, SOLUP, that brings both cheminformatics and first principles-based approaches together in a single platform, where cheminformatics is utilized to improve first principles-based predictions. We will present case studies depicting use of this tool to accelerate the process design and development. In addition to solubility, we will also be talking about case studies where other first-principles based property calculations, such as solvent/API activity, pka, partition coefficient etc., were used to make process decisions . These examples will elucidate the utility of these computational tools to reduce early-stage uncertainty and providing directional knowledge to be used for optimization.

References:

1. Journal of the American Chemical Society 2021 143 (42), 17479-17491
2. Nature Scientific Reports **10**, 11492 (2020)
3. AAPS PharmSciTech 23, 18 (2022)
4. Journal of Cheminformatics 13, 98 (2021)
5. Nature Communications 11, 5753 (2020)

Disclosure: Sharad Maheshwari, Eric J. Murphy, James Marek, Ryan G. Ellis, Akshay Korde, Manish S. Kelkar, Daniel A. Pohlman, Richard S. Hong, Nathan S. Abraham, Rajni M. Bhardwaj, Jeremy Henle and Nandkishor Nere are employees of AbbVie. Kushal Sinha and Moussa Boukerche are former AbbVie employees. All authors may own AbbVie stock. AbbVie sponsored and funded the study; contributed to the design; participated in the collection, analysis, and interpretation of data, and in writing, reviewing, and approval of the final publication.

673d Predicting the Solubility of Amino-Acid Mixtures with the SAFT-γ Mie Group-Contribution Method

Ahmed Alyazidi, Andrew J. Haslam, George Jackson, Amparo Galindo

As peptide drugs gain increasing attention in academic and industrial arenas as active pharmaceutical ingredients (APIs) due to their therapeutic value, more resources are directed to building robust computational models for estimating their solubility in a wide variety of solvent mixtures at variable thermodynamic conditions. Since amino acids are the building blocks of peptide molecules, they provide a key to the structure and, thus, to the modelling of peptides. This role of amino acids as building blocks of peptides can be exploited within a group-contribution (GC) method, such as the SAFT-γ Mie, in which the physico-chemical properties of the fluid are derived from a thermodynamic description between the groups comprising its molecules; the solubility, is one such physico-chemical property that is derivable from the fundamental equations of phase equilibrium. This property plays a critical role in the design, synthesis, and purification of drug molecules.

We use the SAFT-γ Mie [1–4] GC equation of state, in which molecules are treated as heteronuclear chains of fused spherical segments interacting via attractive Mie potentials of variable range, with short-range association sites added to mediate directional interactions. The thermodynamic properties of the molecules are derived from the Helmholtz free-energy expression summing the different contributions (ideal, monomer, chain, and association). An expression for the solubility of the solid solute in the liquid solvent can be derived from the equality of chemical potentials in the solid and liquid phases. The SAFT-γ Mie GC is used to calculate the activity coefficient of the amino acid in solution; this is input to the solubility expression, together with the experimental values of the enthalpy of fusion, the melting temperature, and the change in constant-pressure heat capacity between the solid and liquid phases. The required SAFT-γ Mie group interaction parameters of the amino acids are obtained by adjustment using pure and mixture thermodynamic data of molecules that contain the groups of interest. This SAFT-γ Mie GC approach is found to be very satisfactory for predicting the solubility of various naturally-occurring and synthetic amino acids in pure and mixed solvent environments and in multi-solute mixtures.

[1] Papaioannou, V., Lafitte, T., Avendaño, C., Adjiman, C. S., Jackson, G., Müller, E. A., andamp; Galindo, A. (2014). J. Chem. Phys. 140(5), 054107.

[2] Dufal, S., Papaioannou, V., Sadeqzadeh, M., Pogiatzis, T., Chremos, A., Adjiman, C. S., ... andamp; Galindo, A. (2014). J. Chem. Eng. Data. 59(10), 3272-3288.

[3] Dufal, S., Lafitte, T., Haslam, A. J., Galindo, A., Clark, G. N., Vega, C., andamp; Jackson, G. (2015). Mol. Phys. 113(9-10), 948-984.

[4] Haslam, A. J., González-Pérez, A., Di Lecce, S., Khalit, S. H., Perdomo, F. A., Kournopoulos, S., ... andamp; Galindo, A. (2020). J. Chem. Eng. Data. 65(12), 5862-5890.

673e A Mathematical Model to Predict the Drug Release Profile in a Single-Layered Osmotic Controlled Release Tablet

Bhawana Tomar, Mahesh S. Tirumkudulu, Sweta Manthena, Weili Yu, Alfred Berchielli, Pankaj Doshi

Controlled Release Systems are categorised into four broad drug release mechanisms, namely, Chemically Controlled, Diffusion Controlled, Osmotically Controlled and Swelling-and/or Dissolution controlled. Amongst these mechanisms, the Oral Osmotic pump (OROS) tablet, wherein the drug release is controlled by the osmotic pressure gradient, is often the preferred drug delivery systems due to lower incidence of adverse reactions, better compliance, reduced dosage necessary, independent of the hydrodynamic condition, gastric pH and agitation. Also, a variety of designs are available for such osmotic driven systems, so that a newer delivery design can be developed faster once the release mechanism is clearly understood. Despite the large number of experimental studies on oral osmotic drug delivery systems, few have modelled the details of the drug release process.

In this study, a single-layered osmotic controlled release tablet, which is conventionally also known as an Extrudable Core System (ECS), is modelled in order to accurately predict the drug release rate as a function of several key parameters. These include, the excipients and the active pharmaceutical ingredient, the coating thickness as well as the geometry of the tablet. The model elucidates the key events occurring during the dissolution of a tablet, which is coated with a semi-permeable membrane. These events include, the solvent influx, which is driven by the osmotic pressure differences across the coating, core component (drug, polymer and osmogen) dispersion, tablet swelling due to the solvent accumulation, the hydrostatic pressure build up within, tensile stresses acting on the coating, the extrusion of the dispersed core components and dissolution of the drug particles in bulk. Additionally, the condition for successful entrainment of the drug particles based on the hydrated phase viscosity and other parameters has been derived. The model is validated by comparing the predictions with drug release data for two drugs of differing solubility. The predicted release and the measurements agree well thus confirming the suitability of our model in describing the osmotically controlled drug release process from ECS tablets.

The attached figure compares the predicted drug release profile with measurements, (a) Effect of osmogen concentration on drug release, and (b) Effect of drug concentration on drug release. The solid lines are model predictions while the symbols represent measurements.

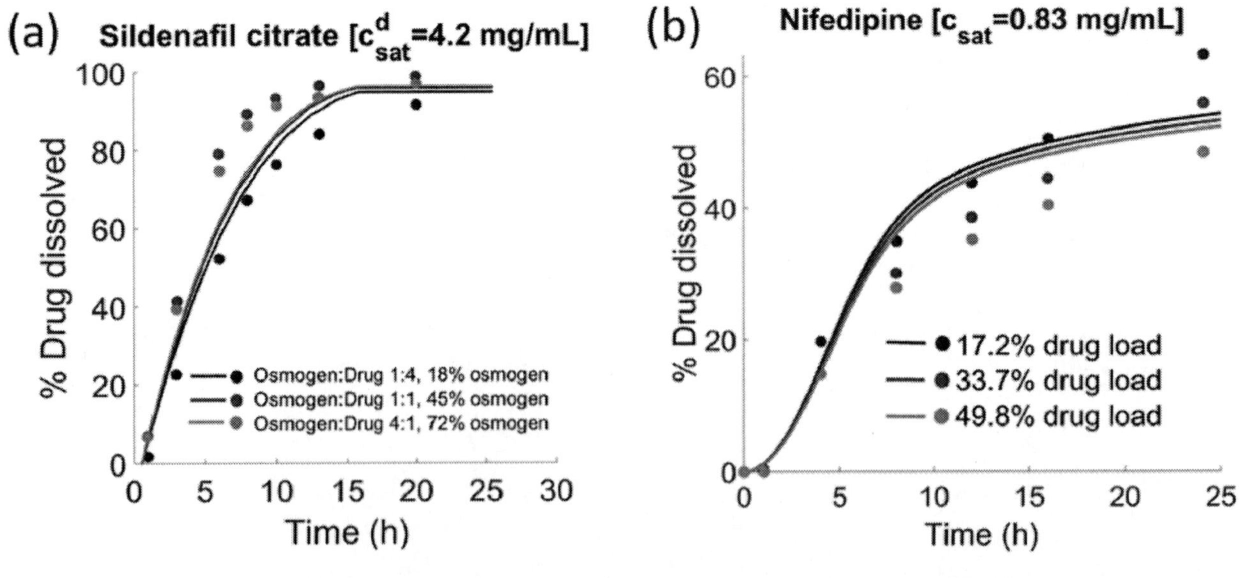

673f Reaction-Diffusion Modeling of Nanocarrier Cocktails for Cancer Therapeutics

Mihalis Kavousanakis, Omkar Bhatavdekar, Stavroula Sofou, Ioannis G. Kevrekidis

Despite the fact that cancer death rates are declining due to early detection, cases of advanced metastatic and/or recurrent solid cancers still have no cure. Current approaches use combinations of therapeutic agents, but unfortunately the vast majority of patients fail to reach a durable response. One of the major reasons for this failure is the uneven distribution of therapeutics within the tumor volume; significant tumor areas are not exposed to therapeutics, therefore their killing action in these areas is limited. In order to address such limitations, we have recently discovered that the delivery of therapeutics is more efficient using two different types of carriers, each preferentially acting on a different region of the tumor. (i) Drug-labeled-antibodies that can effectively kill the tumor perivascular regions and (ii) tumor-responsive liposomes that release their highly-diffusing drug loading in the interstitium, thus enabling the deeper penetration of therapeutics within the tumor.

In order to suggest an optimum combination of the two different carriers we adopt a continuum-level modeling approach that can simulate efficiently the various undergoing processes, namely, carrier transport, carrier binding, drug release and drug transport. Our model is inspired by reaction-diffusion models that have been successfully applied to describe and optimize the performance in heterogeneous catalysis, a classical chemical engineering process that bears great similarities with our system of interest. Solid tumors are composite porous materials hindering the uniform transport of therapeutics throughout their entire volume. The killing action of these therapeutic agents, involve their binding on active sites, and release of their drug-content that will concomitantly kill the cancer cells. Transport and kinetic parameters of our model are estimated by comparing simulations against experimental measurements. Finally, our experimentally informed model can predict the best possible scenario (combinations of the two carriers) for given tumor sizes, target markers expression levels, and cell packing densities by employing model machine learning and data mining techniques.

673g An Adaptive Sampling Surrogate Model for Mixing Time Prediction and Mixing Characterization

Maryam Medghalchi, Saeed Jafari Kang, Iman Mirzaee, Joao de Faria, Fabrice Schlegel, Pablo A. Rolandi

Characterization of process tanks to determine effective mixing and homogenous conditions is a time consuming and cost intensive effort. Therefore, we have developed a comprehensive in-silico model to predict mixing time based on tank specifications, agitation rate and mixture physical properties. Our model is developed based a Computational Fluid Dynamics (CFD) model which is validated in mixing analysis of biopharma mixing vessels. The CFD model is integrated into a four-variables surrogate model in which an adaptive sampling method is applied. Our surrogate model is capable of predicting mixing time and mixing characterization with less than 25% error with only 60 model training samples. The developed surrogate model is available in as a web interface for use by domain experts.

701a Application of Hybrid Modeling Approaches Towards Accelerated Process Characterization

Andrew Fiordalis, Jackie Gonzalez, Karthik Boppidi, Michael Skoumal, Tolutola Oyetunde

The Quality by Design paradigm in pharmaceutical development calls for detailed process understanding that is typically achieved by intensive experimentation especially for poorly understood processes. *In silico* modeling is identified as a potential tool to minimize the cost, labor and time involved in experimentation leading to more efficient use of resources. This goal of reducing the process development cycle has led to intense research efforts in the area of model-based experimental design across the pharmaceutical industry and academia.

However, most of the published model-based experimental design frameworks for biopharmaceutical applications require significant amounts of experimental data as well as considerable technical expertise and/or proprietary software. For early-stage development where process understanding is poor and experimental data are fragmented and limited, it can be challenging to utilize and implement these modeling frameworks.

The case study presented here focuses on identifying experimental design spaces (parameters and ranges) for multivariate process characterization studies utilizing *in silico* models developed using data generated via an initial set of univariate process characterization experiments performed on a CHO perfusion process. The modeling framework is easily transferable to other bioreactor setups and fits into existing experimental process development workflows with minimal modifications. This framework enables smart recommendations for future data collection efforts, iterative model-assisted design of experiments, as well as effective process optimization and control.

In our framework, we utilize mass-balance based multivariate data analytics and gaussian processes to gain initial understanding into the process, calculate specific rates of metabolite consumption and waste generation, as well as assess batch-to-batch variability. We then incorporate information from these first analyses into a hybrid-model structure for prediction of titer and selected key critical quality attributes as a result of changes in process levers. The model also facilitates process optimization and provides insight into areas of the design space that would benefit most from further experimentation (i.e., areas of high variability or poor predictive performance), setting the stage for iterative model-assisted process characterization

With relatively limited experimental data, our framework provides critical insights into factors influencing the process as well as decent predictive capability on several critical quality attributes of interest. We are also able to analyze the impact of different feed schemes, and operating conditions on process performance. Most relevant for process characterization, we use the framework to identify parameter combinations to focus experimental efforts on.

This framework demonstrates the benefit of introducing modeling approaches to in-flight process development efforts with limited experimental data.

701b Risk Mitigation in Model-Based Experiment Design: A Continuous-Effort Approach to Optimal Campaigns

Kennedy Kusumo, Kamal Kuriyan, Shankar Vaidyaraman, Salvador Salvador Garcia, Nilay Shah, Benoit Chachuat

A key challenge in maximizing the effectiveness of model-based design of experiments[1–3] for calibrating nonlinear process models is the inaccurate prediction of information that is afforded by each new experiment[4]. This talk presents a novel methodology to exploit prior probability distributions of model parameter estimates in a bi-objective optimization formulation, where a conditional-value-at-risk[5] criterion is considered alongside an average information criterion. We describe a tractable numerical approach that discretizes the experimental design space and leverages the concept of continuous-effort experimental designs[6,7] in a convex optimization formulation. We demonstrate effectiveness and tractability through three case studies, including the design of dynamic experiments. In one case, the Pareto frontier comprises experimental campaigns that significantly increase the information content in the worst-case scenarios. In another case, the same campaign is proven to be optimal irrespective of the risk attitude. Through this talk, we also introduce a Python implementation PyDEX (https://github.com/KennedyPutraKusumo/pydex) that handles both nominal, risk-averse, and risk-neutral formulations. PyDEX interfaces with convex optimization solvers through cvxpy[8] and can either incorporate user-supplied sensitivities or generate them using numerical differentiation[9].

Available at

Main site : https://github.com/KennedyPutraKusumo/pydex

Stable fork : https://github.com/omega-icl/pydex

References

1. Franceschini, G. andamp; Macchietto, S. Model-based design of experiments for parameter precision: State of the art. *Chem. Eng. Sci.* **63**, 4846–4872 (2008).
2. Atkinson, A. C., Donev, A. N. andamp; Tobias, R. *Optimum experimental designs, with SAS.* (Oxford University Press, 2007).
3. Fedorov, V. V. andamp; Leonov, S. L. *Optimal Design for Nonlinear Response Models*. (CRC Press, 2013, 2014).
4. Asprey, S. P., Macchietto, S. andamp; Pantelides, C. C. Robust Optimal Designs for Dynamic Experiments. *IFAC Proc. Vol.* **33**, 845–850 (2000).
5. Rockafellar, R. T. andamp; Uryasev, S. Optimization of conditional value-at-risk. *J. Risk* **2**, 21–41 (2000).
6. Kusumo, K. P., Kuriyan, K., García-Muñoz, S., Shah, N. andamp; Chachuat, B. Continuous-Effort Approach to Model-Based Experimental Designs. in *31st European Symposium on Computer Aided Process Engineering* (eds. Türkay, M. andamp; Gani, R.) **50**, 867–873 (Elsevier, 2021).
7. Vanaret, C. *et al.* Two-phase approaches to optimal model-based design of experiments: how many experiments and which ones? *Comput. Chem. Eng.* **146**, 107218 (2021).
8. Diamond, S. andamp; Boyd, S. {CVXPY}: {A} {P}ython-embedded modeling language for convex optimization. *J. Mach. Learn. Res.* **17**, 1–5 (2016).

9. Brodtkorb, P. A. andamp; D'Errico, J. numdifftools 0.9.11. (2015).

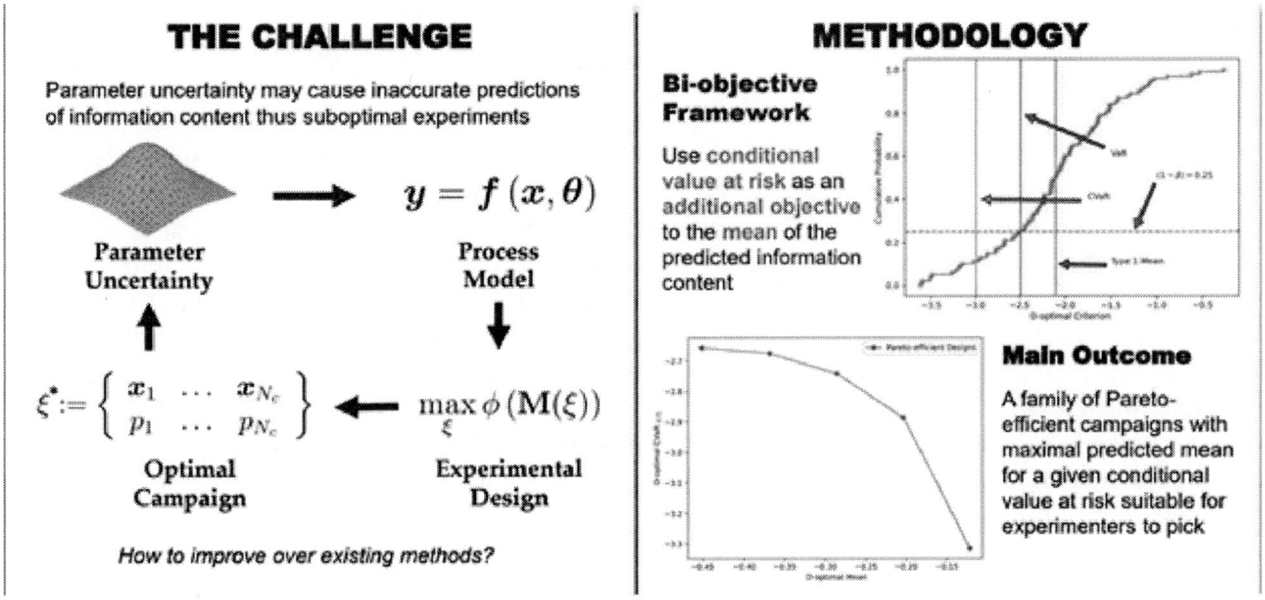

701d Application of a Mechanistic Model for Ultrafiltration and Diafiltration for Process Troubleshooting and Design Space Exploration

Maxwell Maritato, Harun Ozbakir, Xiaoxiang Zhu, Dongying Shen, Joseph Pollastrini, Pablo A. Rolandi

Ultrafiltration and diafiltration (UF/DF) unit operations are a common process step in the generation of a final formulation for monoclonal antibodies. However, UF/DF process dynamics are complex considering factors such as mass transfer, chemical equilibrium, charge interactions, and volume exclusion. Additionally, material for UF/DF experimentation is of high value due to the need for large protein quantities. In silico simulation via a mechanistic model provides the ability to reduce the time, effort, and expense necessary to develop and refine UF/DF operations during process development. This study describes the application of the model developed by Ladwig et al. (2020) for the development of a two-stage diafiltration process. In this work, we describe the use of this high-fidelity model to identify the design space for buffer composition and UF/DF operating parameters in order to generate the desired final formulation conditions. Several in silico design iterations will be presented to describe common questions and elucidate design trade-offs. Sensitivity analysis results will be used to understand model parameters and process variable uncertainties.

References:

Ladwig, JE, Zhu, XX, Rolandi, P, Hart, R, Robinson, J, Rydholm, A. Mechanistic model of pH and excipient concentration during ultrafiltration and diafiltration processes of therapeutic antibodies. Biotechnol Progress. 2020; 36:e2993. https://doi.org/10.1002/btpr.2993

701e Automated Multi-Task Bayesian Optimization of Pharmaceutical Processes

Connor Taylor

In recent years, there has been an increased interest in the use of automated, self-optimizing continuous flow platforms for the optimization of chemical processes.[1] This has arisen from the desire for faster reaction optimization, improved process sustainability and cheaper overall process development. The use of these platforms aims to enhance the capabilities of the researcher by removing the need for labour-intensive experimentation, allowing them to focus on more challenging tasks. These platforms use optimization (machine-learning) algorithms to use only minimal reaction material, but ultimately gain the most process information possible, making their deployment in fine chemical and pharmaceutical industries very attractive.[2]

Recent work has shown that Bayesian optimization in particular is a powerful tool for these applications.[3] However, in all previous studies, each optimization has begun using no *a priori* information about the chemical landscape for the reaction of interest. This protocol therefore requires many initial experimental iterations whereby the algorithm is learning about the experimental design space, without any information on where the optimal reaction conditions may be. This can be wasteful in terms of both cost and time, particularly when there may already be a wealth of information on the chemical transformation of interest from previous optimization campaigns. For medicinal chemistry applications, such as developing chemistries for the elaboration of small polar fragments into nanomolar leads (i.e. Fragment-Based Drug Discovery)[4,5], the use of efficient optimization techniques is paramount due to the minimal quantity and increased preciousness of starting materials.

This work shows the first example of leveraging previous reaction optimization data for unseen chemical transformations, using multi-task Bayesian optimization (MTBO). The framework of MTBO, first introduced by Swerksy *et al.*,[6] replaces the standard probabilistic model in Bayesian optimization with a multi-task model. As these multi-task models can be trained on data from related tasks, we can therefore utilise data from similar reactions - both from the laboratory and from the literature. In this work, we sequentially optimize the reaction conditions for five similar palladium-catalyzed C-H activation reactions using our automated reactor platform. Each reaction yields a pharmaceutically relevant oxindole product, as the optimization efficiency increases with further experimental data - this is true even with different chemical substrates that host different functionalities. One of these five processes is shown in the corresponding scheme, highlighting the reactivity that we targeted in this study.

This work represents a step-change in self-optimization, enabling ultra-efficient reaction optimization through the use of previous experimental data.

References: [1] *React. Chem. Eng.*, 2019, 4, 1545-1554. [2] *Chem. Eng. J.*, 2020, 384, 123340. [3] *Nature*, 2021, 590, 89-96. [4] *Nat. Chem.*, 2009, 1, 187-192. [5] *Chem. Sci.*, 2021, 12, 11976-11985. [6] *NIPS Proc.*, 2013, 2004-2012.

701f Hybrid-Modeling Approaches for Rapid Quantification of Reaction Kinetics

Boung Wook Lee, Anastasia Nikolakopoulou, Mikael Arnold, Xiao Li

Rapid quantification of reaction kinetics is crucial during drug substance process development as key decisions need to be made very quickly. To this end, one of the bottlenecks kineticists face is rapid determination of key reaction kinetics, given the novel nature of active pharmaceutical ingredients and the wide range of the chemistry routes interrogated. This includes kinetic information such as reaction stoichiometry, relative reaction rates, process condition dependency, etc., and potentially complex kinetics mechanisms associated with multi-step elementary reactions, catalyst deactivation, hard-to-detect intermediates, varying reaction time scales, and multiphase reactions. The issue is exacerbated by the fact that different experimental designs are deployed across different organizations, e.g., One-Factor-At a-Time (OFAT) vs multivariate Design of Experiments (DoE), which typically require different data analysis and modeling techniques. This presentation introduces and showcases simple hybrid-modeling approach that can be used to address such challenges. How the introduced approach can complement some of the more recently proposed approaches will also be discussed, i.e., model based DoE, model discrimination, and dynamic DoE.

701g A Fit for Purpose Kinetic Model to Understand the Knowledge Space of a Complex Telescopic Reaction Scheme

Maitraye Sen, Alonso J. Arguelles

The application of kinetic models during chemical development of active pharmaceutical ingredients (API) is vast. However, developing kinetic models, especially for complex synthesis is difficult during early phase pharmaceutical development due to scarcity in data and information. Moreover, precise estimation of kinetic parameters is quite challenging with limited data and resources available. Sen et al. [1] demonstrate that precise parameter estimates are not necessary to develop "fit for purpose" kinetic models that can still guide API process development.

This work will present the development and application of a fit for purpose model to analyze the reaction kinetics of a heavily telescoped complex reaction scheme. The reaction scheme consists of three sub-steps involving several chemical species. A kinetic model structure was established from the proposed chemistry mechanism and information available from similar systems in past. Only certain chemical species of interest were profiled depending on resource availability. Satisfactory estimates of the kinetic parameters were obtained based on the measured responses. This fit for purpose model was used to understand the formation of chemical species of interest as a function of reaction conditions and ultimately to identify a process operating design space at the manufacturing scale.

Reference:

1. Sen et al., React. Chem. Eng., 2021, **6**, 2092-2103

AUTHOR INDEX

AIChE
120 Wall Street, FL 23
New York, NY 10005-4020

CURRAN ASSOCIATES INC.
proceedings.com

ISBN 978-1-7138-7897-1